COLLEGE OF MARIN LIBRARY
KENTFIELD, CALIFORNIA

WITHDRAWN

THE FOUNDERS OF ELECTROCHEMISTRY

SAMUEL RUBEN

DORRANCE & COMPANY
Philadelphia

Copyright © 1975 by Samuel Ruben
All Rights Reserved
ISBN 0-8059-2107-9
Library of Congress Catalog Card Number: 74-20508
Printed in the United States of America

To Rena

Contents

List of Plates vii

Introduction 1

PART I   *The Founders of Electrochemistry* 5

PART II *Count Rumford, née Benjamin Thompson* 11

PART III *Alessandro Volta, Pioneer and Motivator of Electrochemical Evolution* 25

PART IV *Humphry Davy, Innovator in Electrochemistry* 43

PART V *The Motivations of Michael Faraday* 63

PART VI *Reflections* 97

Bibliography 101

Index 103

List of Illustrations

	Page
The Royal Institution in 1840	6
Portrait of Count Rumford	12
Portrait of Alessandro Volta	26
The electrophorus	29
Galvani demonstrating to his family the reactions of a frog's leg when electrically charged	31
Galvani's frog's legs	32
The Volta Pile	36
Volta demonstrating his electric pile to Napoleon	40
Portrait of Sir Humphry Davy	44
Portrait of Michael Faraday	64
An ecological cartoon of 1855	91

Introduction

The decision to write a coordinated review of the founding of electrochemistry and the related lives of the founders came about indirectly as a result of the award to the writer of the Acheson Gold Medal by the Electrochemical Society in 1970. At that time, I was informed that it was a customary procedure to present an acceptance address that could treat a technical subject. However, after giving the matter some thought, I felt that for this occasion it would be of greater interest to present a profile of the life and accomplishments of Michael Faraday, by virtue of whose discoveries electrochemistry became a quantitative science.

From the collection of Faraday literature over the past fifty years, which includes his own diaries, books, letters, and papers given to the Royal Society, as well as biographies by his contemporaries, an address was composed entitled "The Motivations of Michael Faraday." Faraday owed much to his employer and mentor, Humphry Davy, and last year I wrote a paper entitled "Humphry Davy, Innovator in Electrochemistry." Davy's discoveries served as the foundation for applied electrochemistry, and Faraday followed some years later with his discovery of the basic laws of electrochemistry, thereby making it a quantitative science. Each could be classified as a founder of electrochemistry. However, it could also be said that their work was made possible by Alessandro Volta's discovery of the means for generating a steady flow of electricity, a basic requirement for electrochemical studies.

It was Volta's announcement of that epochal discovery to the Royal Society that excited Davy to experiment immediately with this new source of energy. Thus, in relating and correlating the work of the founders of electrochemistry, the life and work of Volta would have to be included.

Since the accomplishments of Davy and Faraday were facilitated by their employment by the Royal Institution of Great

Britain, their use of its laboratory facilities and the encouragement of Count Rumford, Rumford's life and work should also be included as having played an important part in the founding of electrochemistry. His philosophy, when he founded the institution, endowed it with a spirit conducive to imaginative, creative thinking and translation into practical realities. At its founding, he stated that its purpose was "for diffusing the knowledge and facilitating the general introduction of useful mechanical inventions and improvements, for teaching by lectures and experiments the applications of science to the common purpose of life."

His recognition of the genius of Davy and his subsequent employment of Davy was an inestimable contribution to science.

Appreciation is acknowledged for the encouragement given by Professor Henry Linford of Columbia University and Professor Herbert Uhlig of the Massachusetts Institute of Technology to complete this work. I wish to express my appreciation to Dr. Bern Dibner for the privilege of visiting the Burndy Library of Norwalk, Connecticut, which he founded and which contains the most important collections of seventeenth, eighteenth, and nineteenth century books and documents of the history of science, and for the portraits of the founders of electrochemistry utilized in this work. Copies of his writings on Galvani and Volta have been excellent references. The reading of the manuscript and suggestions given for grammatical changes by William Sauerbrey, Leon Robbin, and Wayne Hruden are also greatly appreciated.

The papers on Faraday and Davy were published by the Electrochemical Society and grateful acknowledgement is given for permission to use them in this work.

Samuel Ruben

PART I

The Founders of Electrochemistry

The Royal Institution building in 1840. From a watercolor by H. Hosmer Shepard. Courtesy of Burndy Library, Norwalk, Connecticut.

The early developments in the sciences, and particularly in electrochemistry, were initiated in England during the nineteenth century by men who possessed creative minds and had an inherent sense of direction. They engaged in self-motivated experimental work to fulfill their bold dreams, which in some instances required for practical results the synthesizing of a new technology.

In order to make their experimental efforts more fruitful, they required means for propagating their accumulated knowledge, and this was to be provided by the Royal Institution of Great Britain in London.

The Royal Institution enjoys a special place in the history of science, its early role in the scientific community reflecting the results of an endowed philosophy contributed by its founder, Count Rumford. An adventurous character, he also had a creative mind, and recognized the importance of disseminating knowledge of the latest advances in science in order to improve the nation's arts and industries. Count Rumford had educated himself in the sciences to a degree where he was able not only to contribute to applied science but also to establish the concepts which led to the formulation of the Law of Mechanical Equivalence of Heat. He established an institution that provided an atmosphere conducive to the development of imaginative thinking by gifted individuals, as well as the laboratory facilities necessary for the translation of imaginative concepts into reality. His provision for the presentation of a series of popular lectures and scientific publications helped to relate the laboratory work to industry and obtain support of the institution by an interested public. Fortunately, the successors of Count Rumford included such creative men as Humphry Davy and his successor, Michael Faraday, both of whom achieved greatness while employed by the Royal Institution and contributed to the scientific world the results of their studies and discoveries. They also became interesting and brilliant lecturers who, by their ability clearly to define the nature and advances in science to a fascinated public audience, helped to maintain support of the institution. Throughout its history, the Royal Institution in London has had many famous scholars, but since this narrative relates to the founders of electrochemistry, it will mainly involve the lives and works of Humphry Davy and Michael Faraday.

Such was the Institution's atmosphere at the beginning of the nineteenth century, when the very important discovery of the chemical means for generating a continuous flow of electricity was announced by Alessandro Volta, a professor of physics at the University of Pavia.

The ultimate founding of electrochemistry as a science can be related to the integrated efforts of four men. The first to be considered is the founder of the Royal Institution in London, who provided the opportunity and the encouragement to his employee and successor at the institution, Humphry Davy, whose work in the study of electrical and chemical reactions led to his famous discoveries in applied electrochemistry. Count Rumford, however, was no longer directly associated with the administration of the Royal Institution when Davy discovered, by electrochemical action, the metals potassium and sodium. Nor was Rumford living at the time when Michael Faraday, successor to Davy and director of the laboratory, converted empirical electrochemistry into a quantitative science. Davy was no longer living when Faraday discovered how to generate electric current flow in a conductor by inductive relation with a varying magnetic field. This breakthrough provided the basic principles for the development of dynamos or electric generators of the large amounts of electric currents necessary for the development of industrial electrochemistry. However, it was the spirit and philosophy of the founder of the Royal Institution, Count Rumford, which continued to influence his successors. A profile of the lives and accomplishments of these gifted men of varied personalities will bring into perspective the interrelation of their work which led to the founding of electrochemistry as a science.

Thus, this narration will begin with the life and accomplishments of Count Rumford, the founder of the Royal Institution that became the birthplace of discoveries in electrochemistry.

PART II

Count Rumford (née Benjamin Thompson) Founder of the Royal Institution of Great Britain

Portrait of Count Rumford by Gainsborough. From a copy. Courtesy of the Rumford Historical Association, Woburn, Massachusetts.

Benjamin Thompson, who later became Count Rumford, was born in 1753 in the Town of Woburn, Massachusetts. His family was relatively poor and while still a small child he lost his father. As a consequence, in his early years his mother provided the necessary support. His education was acquired in a small country school with fellow students who were children of the local farmers. In spite of a limited formal education, Thompson had an inherent desire to learn beyond the level provided by local facilities and seriously read from an analytical standpoint, many books relating to subjects that interested him. The combination of an imaginative mind and the ability to interpret the subject matter of these books enabled him to educate himself above the level of his fellow students. He possessed a purposeful and relentless drive to follow through on his studies and made imaginative application of the subjects that interested him most.

At 13 years of age, he had developed a proficiency in arithmetic, handwriting, grammar and spelling, along with a skill in making mechanical devices. He sought for and read technically oriented literature whenever it was available, being particularly interested in the ignition and explosive effects of gun powder. He was apprenticed to John Appleton, an importer in Salem who recognized Thompson's keen mind. During his employment, Thompson collected and studied books relating to ordnance equipment. Another interest was the reading of books describing the romantic grandiose-appearing lifestyle of French upper social circles. He taught himself the French language, which served to intensify his dreams of the "bon vivant" style of living.

In the process of his studies on military equipment, he developed ideas of his own relating to materials. This trend of thinking proved to be of value to him later on in his drive to acquire wealth and social status.

He became knowledgeable and fluent in the subjects that interested him, and did not hesitate to express himself.

His facile mind and inherent interest in reading and study impressed the Reverend Thomas Bernard of Salem, who taught him algebra, geometry and astronomy.

When he was 16 years old he left Salem to be employed in a Boston store owned by Hopestall Capen. He lost interest in commerce and left his job to study medicine with Dr. John Hay of Woburn. He subsequently lost interest in medicine and in 1772 obtained a position as a teacher in a Bradford, Massachusetts, school. This change proved of value to Thompson's education, for besides being a means of support, it brought him in contact with Reverend Samuel Williams, who was impressed with Thompson's keen interest in science and who informally tutored him in a broad spectrum of the physical sciences. Williams, who later became Hollis Professor of Mathematics and Natural Science at Harvard, and founded the University of Vermont, was well qualified to act as a mentor to Thompson.

Thompson, with his friend from Woburn, Loammi Baldwin, also attended a series of science lectures given by Professor John Winthrop at Harvard College. Baldwin has been cited as one of America's first engineers, and his more complete education helped him to assist Thompson toward a better understanding of the natural sciences. Baldwin is also noted for the development of the species of apple bearing his name.

In the summer of 1772, then 19 years of age, Thompson was employed as school master in Concord, New Hampshire. He had developed physically into an attractive young man described as having a manly figure, bright blue eyes and dark auburn hair. While teaching at the Concord school, he met the wealthy widow of Colonel Rolfe and courted her. They were soon married, and this gave him opportunities for fulfilling some of his dreams toward a fanciful mode of living. Thompson had a charming personality and made effective use of it in his social climbing and opportunism. He believed in dressing in clothes that were the most stylish and the best he could afford. He supplemented his good appearance with appropriate grooming and the acquisition of social graces that helped him to enjoy more comfortably the association of his new circle of friends. He devoted his working time to the management of his wife's thousand-acre estate, and through the influence of her family, he gained an introduction to Governor Wentworth. The Governor was impressed with him, recognizing his keen mind and outgoing personality. Thompson's knowledge of military subjects helped him to acquire an appointment as a major in the Second Provincial Regiment of New Hampshire.

Thompson's sympathies and interests were with the Tories during the growth of the American Revolution, and his activities prompted action against him. Learning that the rebels were threatening action against him and accusing him of supplying information to the British in Boston, he fled from his wife's estate to his mother's home in Woburn. When local feelings ran high, after the battles of Lexington and Concord, he was again under suspicion and charged with being inimical to the cause of the revolution. He was questioned by a Committee of Correspondence and managed temporarily to satisfy them and be released. But he was branded as a Tory, and life became very difficult for him. Upon learning of other threats, he fled alone from his home in Woburn to Boston where, safely behind the British lines, he managed to become a member of a group that was returning to England.

Because of his apparent fluency in military matters and his personality, he was chosen by General Gage to be assigned to a commission of four whose task it would be to inform the British officials in London about the local military situation.

In the course of fulfilling his commitments as a member of the reporting commission, he met Lord George Germaine, His Majesty's colonial secretary. Lord Germaine was involved in the military administration that was engaged in fighting the American Revolutionaries. He sensed that this handsome, shrewd individual, knowledgeable in military matters, would be of assistance to him. Thompson advanced rapidly from a clerical position in the colonial office to that of an under-secretary. It has been written that Lord Germaine had a fondness for amorous relations with young men and that Thompson, though not of gay persuasion, would, as an opportunist, go along with any situation that would further his drive toward wealth, fame and the opportunities necessary to pursue his dreams.

While Thompson enjoyed his newly acquired status in his drive for social goals, he did spend time in the development of his ideas concerning military engineering as related to ordnance production problems. He developed a machine for measuring the explosive force of gunpowders. Firing shots into a pendulum, he measured slug momentum and with other experiments, developed an early system of ballistics. In 1781, after a series of experiments on the forces of gunpowders he read his first paper before the Royal Society on this subject. The purpose of this study was to increase the range of balls fired from a cannon. His work was recognized for its scientific worth and in 1779 he was elected Fellow of the Royal Society. The F. R. S. made it easier for him to meet eminent men of this period and increased his perspective for experimental work. On the practical side, in his quest for acquiring wealth, he managed to acquire money while engaged in arms production and amassed a sizeable amount that was dissipated through transitory pleasures in keeping with his desired mode of living.

He had been suspected of being involved in a foreign intrigue, which had it been successful, would have been very profitable, but the plan became known and one of the two conspirators turned State's witness and the other was hanged. Lord Germaine protected his friend Thompson, who apparently knew too much of other intrigues involving his superiors, and sent him to America to fight the colonists as an officer in the Royal Dragoons. It was 161 years before some of his activities recorded in the British archives were opened to the representatives of President

Roosevelt when America turned over to Great Britain 50 over-aged destroyers in 1940.

He returned to England before the war was over and devoted himself to a study of several scientific projects related to munitions production. When the head of the administration resigned at the termination of the war, Thompson decided to visit Austria and Alsace. While observing a military review in Strasbourg, he met Prince Maximilian, who later became Elector of Bavaria.

The Prince urged him to visit his uncle, Karl Theodore, then Elector of Bavaria residing in Munich, and he did. Thompson impressed the Elector, who then invited him to enter the Civil and Military service of the Bavarian government. To accept this offer, he had to have the British government's permission. He returned to England and permission was granted by King George III, who also conferred knighthood upon Thompson.

On his return to Bavaria, Sir Benjamin Thompson was given the rank of Major General of the Cavalry and Privy Counselor of State. During the next several years he introduced many reforms and improvements which were of major importance to Bavaria. He upgraded the Bavarian Army, which was in a demoralized state, by arranging to have the soldiers receive more than the pittance they were given and providing them with proper clothing and better quarters. He instituted new discipline and drills that proved of great benefit to the morale of the army and established a system of public schools in Bavaria, beginning with the children of the soldiers.

To provide the necessary clothing for the Army and to reduce the poverty and mendicancy which were the aftermath of war, he instructed the Army to outlaw begging, and assembled all beggars and unemployed in a House of Industry where they could work for a living by producing uniforms and other military supplies. Those who were accepted for employment were given instruction in various trades and supplied with adequate food and housing. The project was successful and some goods were eventually sold for export. While the program caused the state a sizeable outlay of funds, it eventually became a source of income and ameliorated serious social problems. Another public project which he directed was the reclamation of a large wasteland just outside of Munich. He planned a wooded park with lakes, paths and drives which, when completed, became known as the "English Gardens." The availability of this attractive park to all citizens of Munich was much appreciated and still serves as a recreation area. In order to commemorate his part in this civic improvement, a monument of freestone and marble was erected at the entrance of the garden during his stay in Munich. A medallion likeness of Thompson appears on one side with the following tribute:

> "To him
> Who rooted out the most disgraceful
> public evils
> Idleness and Mendicity
> Who gave to the Poor relief,
> occupation, and good morals,
> and to the Youth of the Fatherland
> so many schools of Instruction.
> Go, Saunterer! and strive to equal him
> In Spirit and Deed
> and us
> in Gratitude."

Fifty-three years after his death, a bronze statue of Thompson, who was then remembered as Count Rumford, was erected by King Maximilian II of Bavaria. It stands on Maximilianstrasse as an expression of respect and gratitude from the citizens of Munich. In 1914, the Rumford Historical Association petitioned King Ludwig III of Bavaria to add the inscription "Geboren zu Woburn, Massachusetts." A full-size replica of this statue was cast in Munich and donated to Woburn by one of its citizens, and now stands on the grounds of the library in his native city.

Thompson was interested in improving the food supply for the army, and as a military official he ordered all soldiers serving the

Duke of Bavaria to plant a patch of potatoes. The early Spanish explorers had learned of potatoes in Peru, and for 250 years they had been considered by Europeans as essentially a cattle food. It was rumored that potatoes caused digestive disturbances in humans, but in France, Monsieur Parmentier, an apothecary, investigated potatoes as a human food and endorsed them with a study of various ways of preparation. Rumford was interested in Parmentier's findings, which prompted the growing and preparation by the army. Potatoes soon spread over the countryside as a valuable food for humans, saving many lives when natural calamities endangered the supply of cereals.

Despite his many civic and philanthropic activities, Thompson managed a personal practical financial participation in the industrial production fostered by him and became quite wealthy during his stay in Bavaria.

The Bavarian government, in appreciation of his services, conferred upon him in 1790 the title "Count of the Holy Roman Empire." He adopted the name of Rumford, the early name of Concord, New Hampshire, and became known thereafter as Count Rumford. He acquired other titles, including Knight of the White Eagle.

Rumford's interest and work in scientific subjects were of importance to him and the results of his studies relating to the physics of heat eventually developed into the basis for a basic law of physics that would indirectly be a tribute to him.

The history of science records several instances where a basic law was derived from analysis, interpretation, and quantification of empirical results. The law of equivalence of energy was announced in the mid-nineteenth century, yet the basic concepts were derived from Rumford's conceptions of the nature of heat that had been translated into experimental proof some fifty years earlier. It began with the observations and experiments he made while engaged in the production of cannon for the Bavarian Army. His work and interpretations were responsible for the end of the Caloric Theory which presumed that a caloric material loss would accompany the production of heat. Rumford was cognizant of the large amount of heat generated in the mechanical operations of boring cannon cylinders. He arranged for some controlled experiments which included the weighing of cylinder and tool parts along with the shavings before and after machining. He also arranged to submerge the tool and machine parts in water and used a blunt tool for maximum frictional effect. He found that after two hours of operation by a rotary boring mechanism driven by two horses that the water temperature rose to the boiling point and remained at a boiling condition only as long as mechanical power was being supplied to the rotary contacting tool. His weight measurements of the machined part and tool plus shavings indicated no overall loss of material and the only other product was the conversion of mechanical energy by friction into heat.

After he had returned to England, he presented, in the Philosophical Transactions of the Royal Society, his celebrated paper "Energy Concerning the Source of Heat Which Is Excited by Friction." He had been earlier recognized by being elected a Fellow of the Royal Society for his scientific work which had appeared in several publications. His demonstration that the product of mechanical energy expenditure was only heat upset the prevailing Caloric Theory, but despite his experimental evidence, he had few converts supporting his deductions. While his observations provided the basis for theories relating to the mechanical equivalence of heat, these ideas were not established until Joule demonstrated that 1390 foot pounds of work would raise one pound of water one degree Celsius and be equivalent to one thermal unit. In May 1842, R. J. Mayer, independently of Joule's work, published a quantitative theory in the *Annalen der Chemie und Pharmacia* that established that there is a simple proportionality between heat and work. He introduced into science the expression "mechanical equivalence of heat."

Among those interested in the problem relating to Rumford's earlier disclosure of the mechanical production of heat was Humphry Davy, who, while employed at the Royal Institution in 1812, demonstrated that when two pieces of ice kept below melting point were rubbed together, the ice would melt at the friction contacting interface, and this occurred even when performed in an evacuated atmosphere.

Rumford also demonstrated that heat would not be conducted through a vacuum, a principle that was later applied to the Thermos bottle. He also explained the heat insulating quality of textiles as being a property associated with the insulating effect of enclosed air cells in the textile. Some practical applications associated with his concepts relating to heat include the drip coffee pot and improved cooking utensils and cooking stoves.

In 1794, after his return to England in good financial circumstances, including a pension from Bavaria, he made a practical demonstration of his support of science by contributing in 1796 £1,000 to the Royal Society and £1,000 to the American Academy of Arts and Sciences for the purpose of awarding a medal every two years for outstanding scientific research. Among those who received the Rumford medals from the Royal Society are Humphry Davy, Michael Faraday, Louis Pasteur, James Clerk Maxwell and John Tyndall. Those who have received the medal award from the American Academy of Arts and Sciences include Josiah Gibbs, Thomas Edison, Albert Michelson, Irving Langmuir, Arthur Compton, Karl Compton and Enrico Fermi.

With the cooperation of Sir Joseph Banks, Count Rumford founded the Royal Institution of Great Britain in 1798. It was his hope that this institution would serve as a medium for translating the advancements in science to the arts and industry and also provide a means for disseminating results obtained from its research and development to an interested and supporting public by lecture series. The Institution policy would be to encourage and support any outstanding talent whose contributions would aid its public support and reputation. It was particularly fortunate to have acquired in the early years of its founding the services of outstanding individuals who would gain international recognition for their academic abilities and their realistic accomplishments. These men, by study, experience and application of their outstanding inherent characteristics, developed into brilliant and popular lecturers capable of relating science in an understandable manner to an admiring and supporting public.

Count Rumford had many friends in the scientific circles of England and had learned from his friend Davies Giddy about the exceptional experimental work of Humphry Davy.

At this period there was much interest and speculation relating to the important discovery of generating electricity by chemical activity as announced on March 10, 1800, by Alessandro Volta in a letter to Sir Joseph Banks, President of the Royal Society. Count Rumford invited Humphry Davy for an interview and learned of Davy's interest and the advances he had made in this new field of knowledge. He was impressed with Davy's understanding of chemistry and his work on galvanism which prompted him to request of the Board of Managers of the Royal Institution that Humphry Davy be appointed assistant lecturer in chemistry. The Board concurred on February 16, 1801, adding other duties such as assistant editor of the journal of the Institute. This employment of Davy and the encouragement afforded him can be considered a very important contribution to science by Count Rumford. It gave the genius of Davy an opportunity to develop rapidly and allow a more effective application of his creative mind. It particularly helped to accelerate the progress of his continuing self-education by his appreciation of library and laboratory facilities, exchange of literature with other science centers and direct communication with members of the Institute versed in the sciences. All these advantages were effectively utilized to produce one of the leading chemical philosophers of his time.

After several years at the Royal Institution, Count Rumford's adventurous character came to activity again and he decided to leave England to live in France, where he could fulfill some of his earlier dreams for the life of a bon vivant. He also wanted to have free time to develop further some of his ideas in physics and to apply the ideas derived from his studies of munition equipment and administration related to his experiences in England and the forming of the Bavarian Military School. He made a number of friends who were in the French military service and his ideas on mobile artillery were reported to Napoleon, who was interested in them, and they were later adopted by the French Military Academy at St. Cyr. The French National Institute elected him a member, which was a rare honor for a foreigner, he being admitted at the same time as Thomas Jefferson.

He fully enjoyed the activities of French social circles, particularly attending and participating in the lively discussions at the salons where were congregated the participants of the sciences and arts.

Count Rumford was invited to the salon of Madame Lavoisier, who was the widow of Antoine Lavoisier, the eminent chemist and acknowledged father of modern chemistry. Madame Lavoisier was wealthy and could readily afford the expenses needed to maintain the salon as a meeting place where savants like LaPlace, LaGrange, Cuvier and Arago could freely talk and dine. It was an attractive activity for her and she participated in it with enjoyment. There was an immediate attraction between the two, and after a period of courtship, Count Rumford and Madame Lavoisier were married in 1805. His attractive appearance and courtly manners, and active interest in group discussions with those engaged in the arts and sciences, made him a popular cohort.

However, while his wife maintained her usual social activities, after a number of years he became less interested in them, having a change of perspective. His desire was now to devote more time to consideration of scientific subjects. Among the contributions derived from such studies was the method of standardizing light by measurement of comparative light intensities. He invented a photometer which consisted of a ground glass plate in front of which was fixed an opaque rod. Comparative measure of light; (for example, a lamp to a candle), was accomplished by altering the position of the lamp until the shadow it cast on the plate was equal in density to that of the candle. He determined the comparative intensity of the lamp to the candle by measuring their respective distances from the plate and expressed it as the square of the difference in distances. Thus, if the lamp distance were three times that of the candle, it was determined to have nine times the illuminating power. He established the term "candlepower" as a unit measurement of light.

In Sir Humphry Davy's *Elements of Chemical Philosophy* he cites the observation of Count Rumford of the relation of light intensity to increased combustion temperature and the effect of closely coupled illuminating wicks as follows.

"Count Rumford showed that the quantity of light emitted by a given portion of inflammable matter in combustion is proportional in some high ratio to the elevation of temperature and that a lamp having many wicks near each other so as to communicate heat, burns with infinitely more brilliance than the lamp now in common use."

Our modern electric incandescent filament bulbs owe their increased efficiency over the single filament units to the use of coiled tungsten filaments that allow mutual heating and radiation effects to produce higher photo efficiency. Rumford also invented a steam radiator as a heating device and arranged for test installations at the Royal Institute and the French Institute.

After a few years, the Rumford marriage began to fail because of a conflict of interests. The Countess wanted to continue the gay exciting life of her social group and still needed the intellectual excitement of the discussions relative to the arts, sciences and philosophies. He, now aging, wanted to have more time for independent scientifically oriented thinking and time for experimental work to prove his theories or ideas. Thus, after four years of marriage, the Count separated from his wife, and according to the terms of the Code Napoleon, received one-half of his wife's estate. He moved to Auteuil where, with his collection of books and apparatus for experimental work, he was able to continue the style of life he now sought.

He died on August 21, 1814, his residual estate being willed to endow a professorship at Harvard University. On his grave in Auteuil there is an elaborate stone dedicated to his memory; however the accomplishment of his founding of the Royal Institution of Great Britain will always be a living monument to him.

PART III

*Alessandro Volta
Pioneer and Motivator of
Electrochemical Evolution*

Portrait of Alessandro Volta. Courtesy of Burndy Library.

A tremendously important discovery of Alessandro Volta provided the inspiration for Davy which led to the latter's innovations and discoveries in electrochemistry, in turn made possible by his effective use of the opportunities afforded him by the Royal Institution. Volta's electrochemical discovery provided the link between the creative men who were inspired by its significance and the results they attained at the laboratories of the Royal Institution.

Up to the time of Volta's accomplishment, electricity had been obtained from electrostatic generators which depended upon friction effects on the surfaces of insulating bodies. The most common type of electric machine utilized, as its potential generating surface, rotating glass discs with friction pads in stationary contact with them. Metal surfaces in contact with these pads were connected to terminals which would allow intermittent discharges of high potential when the discs were rotated. In order to provide a limited storage of the electric force generated, the terminals were connected to a condenser or capacitor which stored the high potential by means of the electrical stress imposed upon the condenser's thin glass separator, achieved by two metal foils serving as charging surfaces. The condensers used at that time were of the Leyden jar type, constructed of a glass bottle or jar with the inner and outer surfaces coated with a thin metal foil. Sufficient space at the top end of the jar was free of coating so as to avoid discharges between the foil ends. While this device increased the total energy per discharge, it was limited to intermittent discharges of short duration in the microsecond range. This apparatus made a satisfactory source of electricity which would help in building up a knowledge of the physics of electrical forces, and also served in this period as an amusement or entertainment device by demonstrating the effect of these high potentials discharging at high frequency through the body without serious effect. Another effect was the observation of the electrostatic fields on the raising of the hair when an insulated person was charged to a high potential.

Volta himself invented a stationary electrostatic generator with a condensing or charge-retaining characteristic: namely, the electrophorus. It was composed of a resin disc having an integral metal contact or plate on one side, which when activated by the rubbing of the exposed resin surface with a piece of dry fur developed a high potential on the resin. By contacting a metallic disc having a glass handle to the electrically charged resin disc, a charge could be inductively extracted and discharged from it. This could be repeated a number of times before the resin was completely depolarized.

Seven years later, Volta invented the condensing electroscope which was an improvement over the ordinary gold leaf instrument in that it could indicate lower potentials. Its basic structure was that of an ordinary electroscope having mounted on its top rod terminal a condenser that had two parts, one of which was permanently fastened to the rod, composed of a flat metal disc having a dielectric layer on it such as thin smooth shellac, and its cooperating electrode, another flat disc of the same diameter supported by an insulating glass handle so that it could be separated from the dielectric-surfaced electrode. By use of a condenser or capacitor in series with the gold leaf electroscope, differences in potential too small to be shown by the unaided electroscope could be indicated.

When the upper capacitor electrode was removed after charging, it caused a rise in potential at the electroscope with a wider deflection of its leaves. Volta's discovery was announced in the form of a letter entitled "On the method of rendering very sensible the weak natural or artificial electricity," and appeared in the Transactions of the Royal Society in London in 1782. While the electrostatic generators, or what were termed electric

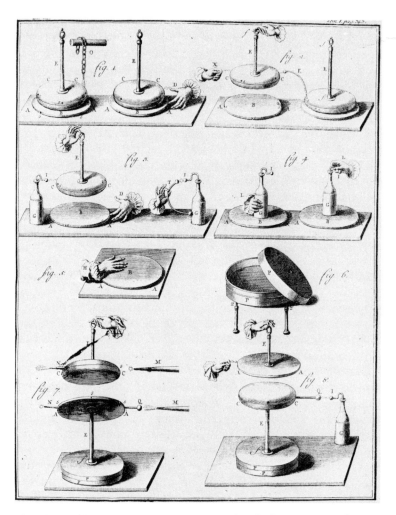

The electrophorus, an early invention of Volta's, was a condenser of a "perpetual" type. Once it was charged, an "endless" number of induced charges was lifted from it, each with some effort. Courtesy of Burndy Library.

"machines," served to provide a means for accumulating early knowledge of electrical forces, they were inherently unsuitable for such applications as electrochemical effects that required a continuous flow of electricity over a sufficient period of time to produce a reasonable amount of end product. The study and application of a continuous flow source of electrical discharge from chemical action, as announced by Volta, changed all of this. The analysis of the development and particularly the determination of the basic chemical factors and quantitative relations responsible for the generation of a continuous flow of electricity by those inspired by his work provided the basis for the founding of electrochemistry.

Volta's work that led to his discovery was initiated and motivated by the announced observations of Professor Luigi Galvani at the University of Bologna. As professor of anatomy, Galvani was occupied in the investigation of the influence of electricity on the nervous reactions of animals, particularly frogs. He observed that when a dead frog was suspended from a copper support which contacted the lumbar nerves, and the crural leg muscles were contacted by the end of an iron wire or arc, the other end of which contacted the copper suspending support, a rapid contraction of the frog's leg occurred.

Galvani had for some time before this studied the effect of electric discharges from electric machines, and had observed induced charge effects on an unconnected scalpel while it was in contact with the nerve center of a frog's leg during the time a spark discharge occurred between the terminals of an electric machine. He also studied the effect of atmospheric electricity and charged Leyden jars on animal organs, and now noted the analogous effect produced by dissimilar metals when a completed electrical circuit existed. He attributed this reaction to the flow of animal electricity and called it the vital fluid from the body to the contacting closed electrical circuit.

Galvani began his study of animal electricity at least a decade before he published his report. He made studies on the marine torpedo, a fish capable of emitting strong electrical discharges.

Professor Luigi Galvani demonstrating to his family the reactions of a frog's leg when electrically charged. Courtesy of Burndy Library.

The contraction of the muscles of other fish that were electrically shocked by contact with the torpedo had been reported by other investigators in 1770. The experiments with the marine torpedo must have been a strong force for influencing his opinion that animal electricity was a vital factor in the muscular contractions he observed in dead frogs.

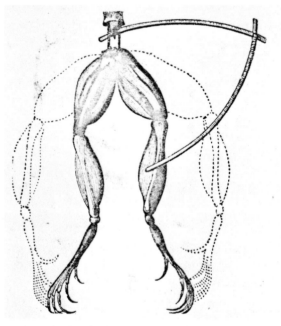

Galvani's frog's legs kicked when a copper rod touched the main nerve and an iron wire touched the rod and a leg nerve at the same time. Volta correctly found the reason for this startling action to lie in a new form of electricity.

Galvani published the results of his discovery in the proceedings of the Bologna Academy of Science in 1791. He sent copies of this paper to his fellow scientists, among whom was Alessandro Volta. His copy is now in the possession of the Burndy Library, Norwalk, Connecticut. Since Galvani's interest was in the physiological effect, he was inclined to the study of the nerve and muscle reactions observed in his discovery of the bi-metal circuit contact effect. Volta, being a physicist, directed his attention to the mechanism involved in the electrical contacting of the nerves and muscles by dissimilar metals.

Galvani's work initiated the term "galvanic," indicating a generation of potential difference between dissimilar materials and "galvanism," which is the art of utilizing these effects.

Alessandro Volta, who was professor of physics at the University of Pavia, was much interested in learning of Galvani's discovery. He communicated with Galvani and after some consideration of the reported results decided to investigate the basic cause of the reactions. He arrived at the conclusion that the energy causing the convulsive effect on the frog's muscle was not the discharge of animal electricity to the contacts, but was essentially due to the electrical force generated by the contact of dissimilar metals separated by the moist flesh of the frog's leg. On May 5, 1792, Volta delivered a paper in the Hall of the University of Pavia in which he discussed Galvani's theory of muscular response to a flow of animal electricity, and his own findings. This caused a strong conflict of opinion between Volta and Galvani, the latter having the support of others on the continent such as von Humbolt. Volta had observed the same reaction that Galvani had noted, that the contraction of the muscles in a dead frog was more energetic when the connecting metal arc contacting the muscles and the lumbar nerves was composed of two different metals. Volta attributed this effect to the contact effect of the metals. He stated that the disengagement of electricity from the contacting metals caused the physiological response, and that the animal served only as a moist conductor to complete the circuit between the dissimilar metal contacts, acting as a sensitive electroscope. He propounded his contact theory with the principle that when two heterogeneous substances are in contact, one of them assumes the positive and the other the negative condition. Volta's conclusions turned out to be correct and it was his many experiments over the years following Galvani's discovery that led him to develop and disclose the means for producing a continuous flow of electricity by chemical action. In an important experiment he arranged a verticle pile of dissimilar metal discs such as zinc and copper or silver, separated from each other by an absorbent paperboard disc that had been saturated with a saline solution. This stack of potential generating couples became known as a "Volta Pile" with its potential or electric force at its outermost terminals dependent upon the number of the series-connected couples. The individual cell potential was determined by the nature of the metal discs and that of the absorbed liquid in the separating spaces. Another form of Volta battery designed to overcome the effects of drying out of moistened paperboard spacers was a circularly placed arrangement of numerous cups filled with various activating liquids such as saline, acid or alkaline solutions. Suspended from opposite sides of the cups' edges were dissimilar metals such as silver and zinc with a series connection between these dissimilar metals for maximum combined potential.

Volta experimented with various metals to determine the maximum electromotive force that could develop between dissimilar materials. He found that the couple which initially had the greatest difference was zinc and graphite, but graphite, when in a pile, did not react with the electrolytes and rapidly polarized. He arranged the elements in a series such as zinc, tin, lead, iron, copper, platinum, gold, silver, carbon and found that the further apart these elements were positioned in this series, the greater was the electromotive force.

In later years this series was greatly expanded by the work of Faraday, Nernst and others, and became known as the electrochemical series. The later arrangement of the series was based on the relation to hydrogen ion electrodes with those elements electropositive to hydrogen on one side and those electronegative on the other. The series affords an effective guide for consideration of voltaic cell compositions and chemical reactions.

Volta may have been influenced in his early work by a publication in Berlin in 1762 by Johann Georg Sulzer of Switzerland when he reported his observation on the contact sensation effect of two dissimilar metals on the tongue. He found that when a strip of silver was placed on the tongue and a strip of lead under the tongue and the two extending strips were contacted together, an acidic taste of iron sulfate was noted. There was no taste sensation until metal contact of the extending metals was made. The electrochemical action causing this effect was not recognized at that time.

This Volta battery was a device that could produce a continuous flow of electricity. The magnitude, time and density of its electrical flow would in time be increased by those who would follow his development. He announced his discovery of the means for producing a continuous flow of electricity in a letter of March 20, 1800, to Sir Joseph Banks, President of the Royal Society in London. The letter was sent from his home in Como and written in French. It had, however, the English title "On the Electricity Excited by Mere Contact of Conducting Substances of Different Kinds." The letter was published in the transactions of the Royal Society. Volta's explanation of the continuous flow of electricity from the pile was based on an action inherent in the dissimilar metals in contact with the moistened spacer. He expected the electric fluid generated to be capable of continuous movement through the complete electric circuit, as can be noted in his letter to Sir Joseph Banks. He called it a "movement perpetual," with the force of the electric fluid increasing with the chemical dissimilarity of the disc electrodes. He sent the letter in two parts, because of the prevailing war, and eventually read it before the Royal Society on June 26th 1800. Volta's reputation was known to the Society, for on April 3rd, 1791, he had been elected a fellow of the Royal Society, and in 1792 he received the Society's Copley Medal for his development of the electrophorus and improved electroscope. In October 1793, Volta submitted a paper for the transactions of the Royal Society entitled "Account

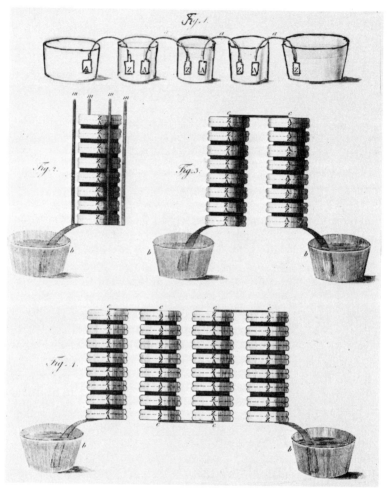

The Volta Pile as depicted in Volta's letter to Sir Joseph Banks. At the top is shown the "crown of cups" and below it variant arrangements of A (silver, for argentium) and Z, zinc discs with moist paper separators. Courtesy of Burndy Library.

of Some Discoveries by Mr. Galvani with Experiments and Observations On Them."

Volta's announcement of March 20, 1800, provided the exciting mental tool for the experimentalists who would repeat his work and themselves contribute to improvements and knowledge through their own findings. Before Volta had the opportunity to read his paper to the Royal Society, there were a number of men inspired by the letter who rapidly engaged in the study of this development. Among those motivated by Volta's announcement were Nicolson and Carlisle, who on May 2, 1800, duplicated the Volta Pile and accidentally discovered that hydrogen and oxygen gases were released from a drop of water separating the two wires from the terminals of the Volta Pile. Cavendish had earlier observed that when a spark was discharged through hydrogen and oxygen they recombined to form water.

Cruikshank, the British chemist, after studying Volta's battery, constructed one of considerably higher capacity. He placed wide copper and zinc plates in a grooved resin-lined wooden box that sealed off the cells, and filled the space between the dissimilar plates with dilute acid or a saline solution. In July, 1800, he was able to obtain a flow of electricity adequate to deposit copper or silver from their solution, thereby pioneering the art of electroplating. The term "cell" applied to Volta batteries was derived from Cruikshank's cell structure. Others such as Ritter, worked on the effect of current discharge through solutions of metal. In September, 1800, he reported his work on electrodeposition of copper from copper sulfate solutions. Volta had used physiological effects such as the observable reactions on muscles or on the organs of hearing, sight, taste and touch to note increases in potential. It was the work of Oersted and that of Ampère who discovered the principles for developing instruments that would instantaneously indicate the voltages or measure the current flow. A physicist, Volta initially approached the problem of generating electricity (as can be noted from the title of his announcement paper "On the Electricity Excited by the Mere Contact of Conducting Substances of Different Kinds") as a physical phenomenon, whereas the chemists who followed his discovery sought the basic chemical cause. The result of the latter approach can be noted in the independent work of Humphry Davy, who proved by Oct. 1800 that the cause was the oxidation of the zinc at the negative pole and the reduction effects at the positive pole. When he used an oxidizing solution in contact with the positive electrode, he obtained a cell having higher potential and capacity.

Working independently of Davy, Fabroni, an Italian contemporary of Volta, had expressed his opinion that the oxidized appearance of the zinc discs contacting the paperboard separator discs moistened with acidulated solution might be the principal factor for the generation of electricity in a Volta Pile.

In recognition of Volta's contribution in presenting to the world its first electric battery, science has honored him for all time by designation of electrical force or potential as the "volt" and cells as "voltaic cells." When later developed, meter devices for measuring the voltage became known as "volt meters." The events in Volta's life that give a better understanding of his accomplishments and world-wide recognition are briefly accounted in the accompanying review.

Alessandro Volta was born in Como, Italy, on February 18th, 1745, to a patrician family. When he was fifteen years old, he was enrolled in a seminary where his studies included the classics and philosophy. He recognized the importance of the knowledge of languages for the better understanding and breadth of his studies. He mastered English, French, and Latin, along with a reading ability in Spanish and Dutch. In later years these studies proved of great value enabling him to correspond meaningfully with savants in several countries. Three years later he became seriously interested in physics and chemistry and engaged in correspondence with a number of authorities in these fields. Among those with whom he corresponded was Professor Beccaria in Turin, on the subject of electricity. He gathered experimental results of his own studies, and in 1771 published a paper, the title of which was "De Vi Attractiva Ignus Electreci."

In 1774, three years after publication of this paper, Volta was elected to the faculty of the Royal School of Como. A year later he was appointed Professor of Physics and at that time he announced the development of a special form of electrostatic high voltage generator known as the electrophorus.

He had developed the ability to translate his concepts into ideas of applied science. For example, in his letter to Professor Bartells, on April 18, 1777, he suggested a signaling system between Como and Milan which would discharge pulses of electric force through a wire between the two cities.

In 1779 he accepted a professorship in physics at the University of Pavia and held this post until he retired.

In 1794 then 49 years old, Volta married Terresa Peregini, daughter of a patrician family, and they had three sons. On November 9, 1793, he lectured at the National Institute of France on electrical effects and Galvani's work, the audience including the French mathematician LaGrange, Count Rumford, Laplace, and Napoleon.

His most important communication was the letter he sent on March 10, 1800, to Sir Joseph Banks, President of the Royal Society, announcing his discovery of the chemical means for generating a continuous flow of electricity.

It can be said that his work initiated the evolution of our electrically oriented society, for his discovery served as a mental and physical tool which resulted in the unfolding of a succeeding and more powerful means of generating electricity. Faraday's great discovery which made this possible was his producing electric current flow induced in electrical conductors by the induction effect of varying magnetic fields. The early magnetic field source was an electromagnet energized by passing electric current from voltaic cells through its windings.

Volta travelled extensively, making voyages to Switzerland, Germany, France, England, Holland and America. He received many honors in recognition of his attainments and was particularly honored by Napoleon, who had attended his lecture at the National Institute of France. Volta addressed the Institute of France in a series of three lectures on November 7, 12, and 22,

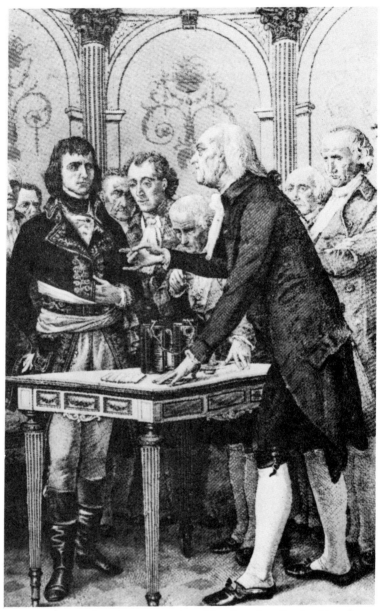

Alessandro Volta demonstrating his electric pile to Napoleon. From a painting by G. Bertini. Courtesy of the Civic Museum, Como, Italy.

1801, which included his work on the electric pile. Napoleon was fascinated by the lecture and demonstration, the importance of which he recognized. He recommended further honors be bestowed upon Volta, and after a commission of distinguished scientists confirmed Volta's discovery, the National Institute Gold Medal was awarded to him. He lectured on galvanism, and illustrated his lectures with experiments such as the decomposition of water by electric current derived from his cells.

Victor Hugo has written that in 1803 Napoleon, while visiting the library of the National Institute, observed on the wall a memorial plaque carrying the inscription "Au Grand Voltaire" which was surrounded by a wreath of gold. He ordered his entourage to have the last three letters (ire) on the plaque removed thus changing it to "Au Grand Volta." The many honors bestowed upon him included Knight Commander of the Legion of Honor, member of the National Institute, Senator of the Realm of Lombardy, the title of Count, and the Italian Order of the Iron Crown." In 1816, he published in Florence five volumes of his work, and in 1819 retired to the family estate near Como. When in Como he met Sir Humphry Davy and his assistant, Michael Faraday, who were on a continental tour of cooperation with other scientists. Volta made a number of improvements on his battery; for example, to restrict the drying out of the moistened cardboard separators between the actual metal discs he suggested ways of sealing the pile by coating it with paraffin or pitch. Volta died in Como on March 5, 1827, at the age of 82.

Fortunately for science, there were many who recognized the depth of Volta's discovery and who assiduously followed their own studies in exposing the exact nature of the chemical process involved in generating electricity. Among the pioneers, Humphry Davy can be cited since he proved the true chemical action in the generating process is one of oxidation and in 1807 was the first to publish a theory of correlating electrical and chemical action. Since Davy was one of the founders of electrochemistry whose dedication to this subject was fostered by knowledge of Volta's work, a profile of his life and accomplishments will be highlighted by his part as Innovator in Electrochemistry.

PART IV

*Humphry Davy
Innovator in Electrochemistry*

Portrait of Sir Humphry Davy. Courtesy of Burndy Library.

Humphry Davy was born in Penzance, Cornwall, on December 17, 1778, the eldest child of a middle class family. He was endowed with a quick, imaginative mind and the ability to learn rapidly from books. These books served not only as a means of instruction, but also as a source of inspiration to him.

He attended a local grammar school where, among other subjects, he was taught Latin and Greek. At fourteen years of age he was enrolled in a better school at Truro where he acquired a knowledge of the classics. Being somewhat a romanticist, he retained this knowledge and in the many lectures which he subsequently gave, he frequently delighted his audiences by quoting related passages from the Greek philosophers. These illuminated his lectures and made them more than briefs on science, thus helping to establish his reputation as a fascinating lecturer and natural philosopher.

After Truro, Davy spent a year applying his imaginative mind to literary fields, particularly poetry. He also engaged in fishing and hunting, activities which, accompanied by an appreciation of the beauty of nature, he loved throughout his life. His ability in composing poetry received, in later years, encouragement from Samuel Taylor Coleridge and Robert Southey, and some of his poems were even published in Southey's Annual Anthology of 1799 and 1800. At this time he had only a speculative interest in the sciences.

When his father died, the family was left with an estate diminished by unwise investments. This situation made it imperative that he acquire a training so that he could assist in the support of the family. He decided that he would like to follow the medical profession. Therefore, on February 10, 1795, he applied for an apprenticeship with Dr. J. Bingham Borlase, a surgeon and apothecary, and was accepted. He enjoyed his duties as an assistant to Dr. Borlase and planned to meet the requirements for practicing medicine by completing his apprenticeship and the necessary studies and taking the examinations at Edinburgh University. He kept an excellent record of his experiences with Borlase, including personal thoughts relative to his work. During his two years of experience with Borlase, he became interested in the chemistry of blood and the problems involving the chemistry of respiration. He would later make some original contributions in these areas.

At this stage in his life, he was fortunate in meeting people who recognized and encouraged his creative thinking, assisting him in the process of self-education and encouraging his interest in the science of chemistry. He had become a good friend of Gregory Watt, a son of the famous inventor, James Watt, who had been educated in the sciences at Glasgow and who admired Davy's keen mind. Watt introduced him to Davies Giddy, a member of the Royal Society, who had a profound influence on the course of Davy's intellectual life. Besides encouraging his creative thinking, Giddy allowed Davy full use of his excellent library, a privilege which was of inestimable value as a motivating factor. In later years, Giddy helped him by introducing Davy's work to Dr. Beddoes and Count Rumford. Giddy changed his name to Gilbert in 1808, and was known by this name in the various offices he occupied in the Royal Society.

By 1797 Davy's interests were principally in the study and understanding of the growing science of chemistry. He was fortunate in becoming interested in it at a time when chemistry was no longer under the influence of the Phlogiston Theory. The work of men in various countries of Europe had changed the thinking from the Phlogiston Theory with its variable unknown "phlogiston" element to a quantitative scientific approach based on measurable changes in weights and volume. This approach was stressed particularly by Lavoisier, the father of modern chemistry, whose measurements with a chemical balance became the true arbiters for determining chemical compositions, reactions, and analyses. The greatest motivating force in 1797, which excited Davy's interest in the study of chemistry, was Lavoisier's *Traité Élémentaire de Chimie*, which gave a comprehensive picture of the science of chemistry as then understood. It contributed to Davy's involvement in a vigorous study of chemistry and experimental work. Lavoisier had the subject of compounds arranged under a system of nomenclature which indicated their compositions, and to some extent, their derivations. Lavoisier also formulated the Law of Conservation of Mass in 1774 upon the basis of experimental measurements, and stated that the total mass of reactants in a chemical reaction is equal to the total mass of the product of reaction. While studying Lavoisier's book, Davy made and collected apparatus to learn the facts of chemistry through practical experience derived from experimental work. There were some items in the book which he could not logically accept and he began gathering the information required to disprove them. His thinking was influenced by the concepts of the Greek philosophers such as Aristotle who stated, 'We need the art of doubting well as the antecedent of progress. Doubt is the parent of inquiry.' Ten years later he was able to prove the fallacy of Lavoisier's description of muriatic acid as "oxymuriatic" acid. He published a paper in which he proved that oxygen was not the essential component of muriatic acid, but that the acid was essentially composed of an aqueous solution of a compound of hydrogen and an element which he identified as chlorine.

One man who assisted Davy by helping him develop a better understanding of the rudiments of science and laboratory equipment was Robert Dunkin. Dunkin was an instrument maker who was also versed in mathematics, electricity and magnetism. He constructed a number of instruments and demonstrated their operations to Davy, thus teaching him some of the rudiments of laboratory science. Davy's highly imaginative, although still undisciplined, mind caused him to express unproven, speculative thoughts as facts in these early stages of his self-education and experimental work. However, bitter experience had taught him how readily one could stray from the truth, and his critical faculties eventually developed, enabling him to engage in meaningful research and to make outstanding contributions to the growing science of chemistry.

He quickly recognized the importance of publishing one's work. Each paper published was like the rung of a ladder—a step toward recognition by one's peers. A faulty rung not based on proven results would cause a halt in progress until it was replaced by one which could withstand the tests of scientific criticism.

Davy's two-year apprenticeship to Borlase had inspired an interest in the chemistry of physiology, and particularly in the chemistry of respiration in a wide range of living organisms. As a consequence of his studies in these areas, he was able to prove that the carbon cycle operated in the sea as well as on land. Davy was fascinated by a book written in 1795 by Dr. Mitchill entitled *Remarks on the Gaseous Oxyde of Azote and Its Effects*, and in 1798, became critical of Dr. Mitchill's statement that nitrous oxide gas was an agency for spreading contagious diseases, producing ill effects when it was breathed in or came in contact with the skin. Davy set up a simple assembly of apparatus necessary to produce nitrous oxide. He then tested the effects of the nitrous oxide gas on animals by having them breathe it or by allowing skin contact. Davy found that the animals experienced no ill effects. He experimented with it on himself and found some very interesting results unrelated to the statements published in Dr. Mitchill's book.

Since he needed large quantities of nitrous oxide gas to continue his research on its physiological effects, he wrote to Dr. Beddoes at Bristol about his experiments. Dr. Beddoes knew of Davy through their mutual friend, Davies Giddy, and on October 2, 1798, offered Davy the position of chemical assistant at the Medical Pneumatic Institute. This institute, founded by Dr. Beddoes, was dedicated to the study and application of the therapeutic value of various gases. Davy was asked to complete an account of his initial work on nitrous oxide and this was published in 1799. A year later he published an intensive study on the effects of nitrous oxide gas on humans in a separate volume entitled *Researches, Chemical and Philosophical, Chiefly Concerning Nitrous Oxide*. He reported pleasurable effects upon breathing the gas and found it was not toxic. Also, he found that it had an analgesic effect in small quantities, relieving him of a toothache which returned when the gas was discontinued. He discovered that it had an anesthetic action when taken in large quantities. Thus began the history of what in dental practice has been referred to as "laughing gas." Davy found that some of the gas was absorbed in the bloodstream, but that it did not combine chemically with the blood and could be freely eliminated without having a toxic effect. Further studies of the physiological effects of other gases, such as nitric oxide and a mixture of carbon monoxide and hydrogen, nearly proved fatal to him. In his work on gases, he had determined the residual capacity of the lung. The publication of his work in 1800 created great interest and was a true rung in his ladder toward professional recognition and fame.

In the Spring of 1800, he learned of a great discovery; chemically produced electricity. This discovery would play an important part in his rise to fame since it helped to make possible his notable contributions to electrochemistry. Davy recognized the importance of Alessandro Volta's announcement of the electric pile which had been described in a communication to the Royal Society. Volta's electric pile, composed of dissimilar metals separated by an absorbent spacer saturated with an aqueous conductive solution, for the first time made possible the generation of a continuous flow of electric current. Prior to this time experiments with electricity were limited to intermittent discharges of short duration provided by an electrostatic generator and stored in a Leyden jar. Now, electrical effects could be studied with a continuous controllable flow of current which was more readily measurable with regard to intensity and quantity. Utilizing the electricity from voltaic cells, Davy duplicated the work of Nicolson and Carlyle who had reported the fact that when an electrical current was passed through water, hydrogen and oxygen were liberated at their respective poles. Davy believed there was much to be learned about the fundamentals of voltaic generation of electric currents. He duplicated the higher capacity voltaic cell structure of Cruickshank and became involved in a study of the controlling chemical factors. When he constructed a voltaic cell using zinc and silver electrode elements separated by a porous spacer containing muriatic acid, he obtained greater electrical force than with electrodes separated by spacers impregnated with saline solutions. He reported that cells having spacers containing nitrous acid were capable of even greater electric output. With this work he proved that the generation of electric current was dependent upon the oxidation of its zinc element and that the total force generated was the result of the chemical activity of the solution in contact with the positive electrode. He investigated many other compositions of solutions in contact with dissimilar metal electrodes and related their composition to voltaic cell characteristics. In his work on various active solutions, he made dual-electrolyte cells with the more highly oxidizing electrolyte in contact with the positive electrode. Davy found that carbon derived from charcoal could be used with zinc to form a spaced couple which would generate an electric current. This finding was not considered to be of great importance because Volta had experimented with carbon electrodes, but years later in the development of commercial electric batteries such as the LeClanché cell and its progeny, the dry cell, carbon became the electrode material for contacting the negative reactant or depolarizer such as manganese dioxide.

In September, 1800, Davy's first paper on galvanism was published in *Nicolson's Journal*. A second paper followed three weeks later. In early 1801 Count Rumford, who was acquainted with Davy's work on galvanism, invited him for an interview. This led to a recommendation to the Board of Managers of the Royal Institution that Davy be appointed assistant lecturer in chemistry. On February 16, 1801, the Board concurred, adding other duties such as assistant editor of the Journal of the Institution.

Davy continued his work on galvanism after his employment at the Royal Institution, and six weeks later gave his first lecture on this subject. This lecture, on April 25, 1801, was attended by Count Rumford, Sir Joseph Banks and other distinguished philosophers. It was a brilliant success and was reported in the *Philosophical Magazine*. Five months later he was promoted to Lecturer in Chemistry.

In January, 1802, his unique personality was expressed when he gave his first course in chemistry. His highly imaginative mind aided him in expressing his ideas concerning chemistry and relating chemistry to all the sciences. His audience was fascinated by the breadth of his vision in relating chemistry to mineralogy, biology, zoology, physiology, and medicine. He philosophized on the opportunities presented to those engaged in related disciplines through a better understanding of chemistry. These lectures indicated the thorough knowledge of the sciences he had acquired from literature, through discussions, and by keen observations in his experimental work.

Count Rumford had originally intended that the Royal Institution orient its studies toward industrial problems, and in order to increase its financial support, it began to do so. The important English industries, dyeing and tanning, felt a great need for better processes, and Davy was asked to interrupt his laboratory work on galvanism to study these needs. After two months of study and visits to tanneries, Davy returned to his laboratory to investigate new and old substances useful for tanning and a method of depilating skins with alkaline materials. His observations and comments on the processes of tanning were published as a tanner's guidebook which was used by the industry for many years. In 1805 he was awarded the Copley medal by the Royal Society for his work on the chemistry of tanning.

The success of Davy's lectures, largely due to his humanistic approach to an interpretation of the sciences, had a secondary effect of materially helping the Royal Institution in a period of financial stress. This was appreciated by the Board of Managers and on May 21, 1802, they conferred upon him the title of

Professor of Chemistry. On November 17, 1803, he was elected Fellow of the Royal Society. The Board of Managers also extended his duties by requesting him to plan and give a course on agricultural chemistry to the members of the Agriculture Board. These lectures were based on the results of many experiments, discussions with farmers and extensive study of the available literature on the subject of agricultural chemistry.

In December 1803, the renowned chemist, John Dalton, delivered a course of lectures at the Royal Institution. He had published his book on a new system of chemical philosophy containing his atomic theory. Dalton's theory postulated the existence of minute individual particles or atoms of different elements having different atomic weights. Compounds were the result of the union of dissimilar atoms, the ratio of whose weights was proportional to the combining weights, combining to form chemical compounds. This accounted for the facts of chemical combinations now known. Dalton and Davy mutually gained from their discussions of their respective lectures and of Dalton's atomic theory of chemical combination. Davy rejected Dalton's atomic theory when he later studied electrochemical conduction, adopting a theory provided by Riggio Boscovitch, an 18th century Jesuit priest, who postulated that atoms were immaterial mathematical point-centers of force surrounded by shells of attraction and repulsion. In 1804 Davy was elected a Fellow Commoner of Jesus College, Cambridge. In 1805 he was asked to collect minerals in Cornwall, the lakes in Wales and in Ireland, and to study the geology of these regions. Davy made an analytical study of what he observed and cast it in a comprehensive report. Later, he gave a lecture on this subject. Continuous study and laboratory work accompanied by creative thinking helped establish his reputation as one of the outstanding philosophers of chemistry.

Davy was now allowed to resume his experimental work on the studies of galvanism. He showed that the relation of electricity to chemical affinity was basic to the understanding of electrochemical action. Using the electricity from voltaic cells, he was able to prove that one could change the material state of substances and their residual chemical properties.

Davy's interests and thoughts in philosophy were influenced by such men as Dematoritus (5th century B.C.) and such philosophers as René Descartes, both of whom expressed the thought that the way to a real knowledge of the nature of materials was to separate substances into their ultimate components. To Davy this concept could be fulfilled by the electrochemical separation of compounds, as referred to in his lectures. His first Bakerian lecture before the Royal Society on November 20, 1806 entitled, "On Some Chemical Agencies of Electricity," was principally devoted to electrochemical decomposition. Davy predicted the value of electricity in discovering the true elements of materials. He proved this later by his isolation of metallic potassium and sodium by electrochemical decomposition of their hydroxide compounds.

One of his important contributions toward the establishment of the fundamentals of electrochemistry was his generalization on the study of migration of materials in solution caused by the application of an electrical current. He stated that,

> Hydrogen, the alkaline substances of metals and certain metallic oxides are attracted by negatively electrified metallic surfaces and repelled by positively electrified metallic surfaces and contrawise, oxygen and acid substances are attracted by positively electrified surfaces; and these attractive and repulsive forces are sufficiently energetic to destroy or suspend the usual operation of chemical affinity. It is very natural to suppose that these excellent and attractive energies are communicated from one particle to another particle of the same kind so as to establish a conducting chain in the liquid and that the locomotion takes place in consequence.

Considering the state of the art at this period, Davy's explanation of chemical and electrical affinities was an instructive guide for work in the developing science of electrochemistry. His imaginative mind foresaw the industrial possibilities of using electricity to decompose neutral materials and, thus, to produce alkaline or acidic compounds on a large scale. His lecture, "On Some Chemical Agencies of Electricity," created much interest for it brought a new concept to the minds of chemists. In 1807 he was awarded the medal and prize from the French Institute for the best experimental work of the year on galvanism. Napoleon had established this medal when he was First Consul, and it was awarded to Davy despite the fact that England and France were at war. Between 1802 and 1806 Berzelius of Gotland, Sweden, had published a number of papers on his electrochemical researches. These undoubtedly helped Davy in laying the foundation for his important discoveries.

Davy's work on the electrochemical decomposition of fixed alkalies led to his most famous discovery on October 6, 1807—the isolation of the metals potassium and sodium. He accomplished this by the electrochemical decomposition of molten potash which produced metallic potassium for the first time. The same technique was later used to produce metallic sodium. He initially named the metals "potagen" and "sodagen," but later changed them to "potassium" and "sodium." He not only discovered these metals in a pure state, but also determined their physical and chemical properties. At a later time he used the same methods in an attempt to produce the alkaline-earth metals, but was not successful. This failure was due to the fact that hydroxide compounds of the alkaline-earth metals are not as electrolytically conductive or sufficiently fluid when heated. By using the method reported by Berzelius; namely, that of using a mercury electrode to form amalgams, he was able to produce some of the alkaline-earth metals. Davy also produced alkaline-earth metals and boron by reduction of their oxides with the alkali metals. His work in proving that the strongest alkalies were compounds of oxygen and the alkali met with great opposition in France. Davy's work was castigated, for it upset Lavoisier's concept that the oxygen was the acidifying element. His discoveries supported his own theories concerning the electrical nature of chemical affinity—a fundamental concept of electrochemistry.

In December, 1810, Davy travelled to Dublin where he gave a series of lectures on chemistry and applied electrochemistry which were continued in the winter of 1811.

In 1811 the degree of Doctor of Laws was conferred upon him by the Provost and fellows of Trinity College, Dublin.

Davy published two important books, *Elements of Chemical Philosophy* in 1812 and *Elements of Agricultural Chemistry*, a year later. His Elements of Chemical Philosophy, noted for its excellent presentation of contemporary chemical science, included a sketch of the history of chemistry. Davy's wide reading enabled him to relate the work of the alchemists and the chemists of Lavoisier's period to the development of the science of chemistry. He stressed his thoughts on the laws governing chemical combinations and proportions in which elements combine. The imaginative thinking of Davy can be noted from the statements in the following paragraph,

> ...whether matter consists of individual corpuscles or physical points endowed with attraction and repulsion, still the same conclusion may be formed concerning the powers by which they act and the quantities which they combine and these powers seem capable of being measured by their electrical relation and the quantities on which they act being expressed by numbers. The laws of crystallization of different proportions and the electrical polarities of bodies seem to be intimately related and completely illustrative of their connection and probably will constitute the mature age of chemistry.

Many years later the prophetic meaning of these sentences was appreciated.

On April 8, 1812, he was knighted by the Prince Regent and shortly thereafter married Jan Apreece, a wealthy widow well known in social and literary circles of England and Scotland. She was status-minded and very anxious that he devote more time to social and literary activities, which were somewhat in conflict with his own inner persuasion toward experimental science. While this fulfilled his cultural interests, it must have had a restricting effect on the maximum expression of Davy's creativity in science.

He admired her very much in the early stages of their union, as can be noted from the following dedication to her of his *Elements of Chemical Philosophy* in 1812.

> There is no individual to whom I can with so much propriety or so much pleasure dedicate this work [as] to you. The interest you have taken in the progress of it has been a constant motive for my exertions, and it was begun and finished in a period of my life [which] owing to you has been the happiest. Regard it as a pledge that I shall continue and pursue Science with an unabated ardour. Receive it as a proof of my ardent affection which must be unalterable for it is founded upon the admiration of your moral and intellectual qualities.
>
> H. Davy

Unfortunately, throughout their lives the record indicates a non-fulfillment of a dream in their married years.

He returned to the laboratory to continue his experimental work and investigated the properties of a compound of nitrogen and chlorine discovered by Dulong in Paris which had aroused his interest. While performing experiments to determine the compound's characteristics, a vial exploded and wounded him by blasting small splinters of glass into his face. He concluded that the compound was nitrogen chloride and presented a paper on its properties to the Royal Society. Davy discovered that chlorine supported combustion and later found the relation of this property to other halogens.

During Davy's last four lectures on chemistry at the Royal Institution, there was a young man in the audience who was enthralled with the lectures and wrote a meticulous set of notes.

Michael Faraday had been invited to the lectures by Mr. Dance, a member of the Institution and an admirer of Faraday's keen mind and serious dedication to studies of the chemical sciences. Faraday sent Davy a plea for a position as a chemical assistant at the Royal Institution and included the notes he had made of Davy's lectures. Davy recognized the unique qualities of Faraday's mind, but was unable to employ him as his chemical assistant at that time. He promised to communicate with him if such an opportunity occurred. He did give him a few days' work as his temporary secretary while his eyes were recovering from the effect of the nitrogen chloride explosion. Three months later Davy discharged his laboratory assistant for misconduct and recommended to the Board of Managers that Faraday be engaged as his chemical assistant. They concurred on March 1, 1813. This appointment proved to be of tremendous benefit both to the Royal Institution and to Faraday. The Royal Institution benefited from Faraday's skill in measurements, his patience and his unerring intuition; Faraday benefited from a close association with the outstanding chemical philosopher, Humphry Davy.

Davy and Faraday had certain qualities and interests in common. Both had creative minds and were compulsive readers of scientific literature. They both educated themselves to be experts in their subjects. It has been written that of the many outstanding contributions Davy made to the Royal Institution and to science, his recognition of Faraday's abilities was one of his greatest.

Davy asked Faraday to accompany him as his assistant on a trip to the Continent in October, 1813, about six months after Faraday had become his chemical assistant. Lady Davy was to accompany them as well. There they would meet prominent scientists such as Ampère, Laplace, Gay-Lussac, Count Rumford, Chevreul, Berthollet, Guyton de Morveau, Humboldt, De la Rive and Volta for discussions and exchange of information. They would also study the volcanoes of France and Italy. They received passports and *laissez-faire* from France, even though England and France were still at war. In their private coach, they carried the portable chemical apparatus which they used in demonstrating chemical experiments.

Shortly after his return from the trip to the Continent, Davy devoted study time to the problem that had plagued the coal mining industry: the recurrent explosions in coal mines caused by the ignition of coal gas and air. Many lives had been lost over the years in all countries by miners caught in these explosions which were ignited by their illuminating equipment. Davy's studies on combustion and heat transmission through confined areas led to his invention, in October 1815, of the mine safety lamp in which the flame was enclosed by copper gauze. This lamp could, therefore, be used in atmospheres containing explosive quantities of the coal gas without igniting them. It also gave a qualitative indication of the presence of explosive concentrations of combustible gas by increases in the intensity of the flame. A flame of decreased intensity was an indication that the oxygen content in the atmosphere was less than 18%. While records show that others had worked on a mine safety lamp, it was Davy's systematic experiments on utilizing the heat conductivity of surrounding wire gauze cylinders for the prevention of an ignition path from the flame of the wick to the external atmosphere that made his lamp the first effective and practical mine safety lamp.

Davy believed that the duty of men of science was to contribute their discoveries to the benefit of mankind. He refused to patent his lamp and dedicated its use to the mining industry. In October 1818, he received both the Rumford gold and silver medals from the Royal Society for his development of the safety lamp. During that same year the Baronetcy was bestowed on him. He was also presented a service of silver plates by the Newcastle mining industry. In his will he requested that these be used for the establishment of the annual Davy medal for outstanding discoveries in chemistry.

In 1818 he went again to Italy with a plan to devise a method of unrolling the papyri documents discovered at Herculeum. On November 30, 1820, he succeeded Sir Joseph Banks as President of the Royal Society, a position he held until 1827.

Davy became interested in electrical and related magnetic effects. He had been impressed by the announcement of Oersted's discovery of the magnetic properties of a conductor carrying a current. He was very interested in the announcement of Ampère that had extended Oersted's discovery in what is now known as "Ampère's Rule" for the relative direction of current flow and the direction of deflection of magnetic needles. Ampère had discovered the mutual force action of two parallel current conducting wires on each other and the magnetic fields of helices with their ability to magnetize soft iron. In investigating the magnetic effects surrounding an electrically heated platinum wire, Davy found these effects to be independent of the wire's temperature. He reported experiments in which steel needles placed transversely to conductors became magnetized. In basic observations on the conductivity of different metallic conductors, he showed that the melting point limited the quantity of electricity that could be transmitted through the conductor and that its conductivity was increased if cooled, and the converse occurred when it was heated.

In January, 1824, he invented a method of galvanic protection of the copper sheathing of the hulls of naval vessels by electrically connecting a mass of more oxidizable metals such as iron, tin or zinc to it. This method, which had worked in the laboratory, failed when tried in actual sea water. This was not because of incorrect reasoning, but because of the rapid overlying accumulation of organic materials such as weeds and barnacles which obscured the galvanic protective action. A century later this process was to prove invaluable in the protection of oxidizable metal structures. Davy demonstrated the feasibility of using electricity as a source of light by connecting the terminal of his large battery to charcoal rods and obtained a brilliant arc light. He also used the electric arc as a furnace element with which he reduced diamonds to graphite and was able to melt many refractory substances.

His Bakerian lecture on October 1824, on the relation of

electrical and chemical changes, contained his last published thoughts on electrochemistry and earned him the Royal Society Medal.

The minutes of the Board Managers of the Royal Institution of February 4, 1825, include the following item.

> It appears that Sir Humphry Davy, having stated that he considered the talents and services of Mr. Faraday, assistant in the laboratory, entitled to some mark of approbation from the managers, and these sentiments met the cordial concurrence of the board; Resolved, that Mr. Faraday be appointed director of the laboratory.

Thus, Michael Faraday, the man whom Davy had served as mentor and whose career had taken on significance as Davy's laboratory assistant, now, twelve years later, became the director of the laboratory. That Faraday in the ensuing years fulfilled his sponsor's expectations was evidenced by his many discoveries. He was the first scientist to place electrochemistry on a quantitative basis and the first to generate an electric current flow in a conductor by induction from a magnetic field.

Davy was elected President of the Royal Society for the seventh time in 1826. But in that year he suffered a stroke and resigned from the Society. His health was failing rapidly and he decided to leave England and settle in Rome. In 1828, having to forego scientific work and field sports, he devoted his time to writing a book on fishing, *Salmonia*, which he illustrated with his own drawings. Although partially paralyzed from several strokes, he spent his last month in Italy writing a series of dialogues which were posthumously published under the titles *Consolations in Travel* and *The Last Days of a Philosopher*. He suffered another stroke and was soon joined in Rome by his wife and brother John. His recovery was slow, and he insisted on moving to Geneva where he died during the night of his arrival May 29, 1829, at the age of fifty.

In generalizing on the self-motivation of Sir Humphry Davy— a characteristic which helped to produce one of the outstanding scientists of his era—the following may be cited: his love and respect for books; his imaginative mind which gave expression to new concepts and translated them into reality; the inspiration and encouragement derived from people with whom he shared common interests; public interest and, finally, the encouragement given by his peers to the establishment of electrochemistry as a science.

For Davy, the study of chemistry was also an aesthetic experience, for he stated that "the study of nature must always be more or less connected with the love of the beautiful and the sublime, it is eminently calculated to gratify the more powerful passions and ambitions of the soul, it may become a source of consolation and happiness." He found an outlet for his philosophy in his poetry and lectures. His great popularity with his audiences was in recognition of the communication to them of his philosophy.

The effective translation of imaginative thinking into practical realities requires the acquisition of knowledge relative to existing and past technology. By continuous study of the facts of science plus experimental experience, a broad mental tool storage can be attained which can be retrieved and coordinated for a harmonious solution of a problem. This process involves in many instances the need for synthesizing a new technology.

Davy was fortunate to have Alessandro Volta's epochal discovery as a source of inspiration and innovation. His vision of an industrial electrochemical processing was fulfilled by those who followed him. They were able to do so with the application of adequate power generated by machines that utilized the principle of magnetic induction discovered by his protegé, Michael Faraday.

He was torn in spirit between two loves—creative science and poetic expression. His friend, Wordsworth, the renowned poet, commented on Davy's integration of the study of chemistry with aesthetic satisfaction in his *Lyrical Ballads* (1802) when in the preface he wrote:

> Poetry is the first and last of all knowledge—it is as immortal as the heart of man. If the labours of Men of Science should

ever create any material revolution, direct or indirect, in our condition and in the impressions which we habitually received, the poet will sleep then no more than at present, but he will be ready to follow the steps of the Man of Science. The remotest discoveries of the chemist, the botanist, the mineralogist will be a proper object of the poet's art upon which it can be employed.

A review of Davy's work and life illustrates how important is recognition and encouragement combined with an opportunity for carrying out conceptual thinking.

PART V

The Motivations of Michael Faraday

Portrait of Michael Faraday by H. W. Pickergil.

Michael Faraday was one of the greatest scientists and yet most humble characters to grace the realm of science. He lived in an era when the accomplishments of an individual were limited by birth or social status.

What philosophy of life and what actions brought this informally educated poor boy to the pinnacle of success and fame? It may be stated that essentially three outstanding factors were responsible for this accomplishment. The first was his unique character which gave him the courage and persistence for converting his imaginative thinking into realistic meaning with a dedicated honesty to himself as well as to others. Secondly, one must cite his appreciation of books together with his early recognition of their teachings for discipline; namely, methods and true interpretations. Finally, the third basic factor was his contact with people who recognized his outstanding qualities and assisted him in his process of self-education and thus encouraged his gifted thinking.

Faraday himself, in one of his letters, described his early education as being most ordinary, consisting of little more than the rudiments of reading, writing and arithmetic as taught in a common day school. However, he was born with a compulsive desire to acquire knowledge which developed into a life-long dedication to the understanding of science.

Michael Faraday was born in 1791 in Newington, Surrey, the third in a family of four children. His father, a blacksmith, died just as Faraday was on the road to a practical scientific education and his desired occupation. His mother was fortunate to live long enough to witness the rise of her devoted son from a mere errand boy to the office of Director of the Royal Institution, the first

Fullerian Professor of Chemistry, and to be acclaimed by the world for his profound and significant discoveries. Her great pride in him was expressed by always referring to him in her letters and conversations as "My Michael."

Throughout his life, Michael Faraday was an inherently religious man and this fact was of great influence in the development of his character. He consistently followed the teachings of the Sandamanian Church, a small dissident Presbyterian sect, which his parents had attended. The creed of the members of this church was that they did not believe themselves to be subject to any league or covenant, but were governed only by the doctrines of Christ and his apostles. They held that Christianity never could be the established religion of any nation or power without becoming the reverse of that for which it was instituted. Furthermore, the Sandamanian Church held that the Bible was the revelation of the will of God in both the inspired and written word. In his later years, when he was elected as elder, Faraday preached to the assembled congregation on several Sundays. To him, his religion was a living root of fresh humility and from first to last, a philosophy which nourished his inner strength.

When he was thirteen years old, the time had come for Faraday to acquire a trade so that he could have a means of support. A short distance from where the Faradays lived was a book shop owned by Mr. George Riebau. Here, young Faraday was hired as an errand boy whose duties were to deliver and collect papers from Mr. Riebau's clients. One of the most important events in Faraday's life occurred when he was indentured by Mr. Riebau a year later as an apprentice to learn the trade of book binding. This was a prime stimulus in his career. Faraday, at a later date, wrote "Whilst an apprentice I loved to read the scientific books which were under my hand."

Books were the inspiration as well as the instruction for Faraday. Three particular books had a profound effect on the channeling of Faraday's inherent mental capacity toward an orderly and coordinated accumulation of knowledge in the field of science. The most important one, providing a guidance which he followed throughout his life, was entitled, *Improvement of the Mind*, written by Dr. Isaac Watt. It taught Faraday the correct process of studying, of lectures, the importance of critical discussion with others and direct observation as a means of acquisition of knowledge with caution, to avoid adopting a theory from a few experiments. Watt stressed the importance of precise language and the necessity of avoiding hasty generalizations. He advised keeping written notes on all ideas and subjects of interest for future reference. The second book, Mrs. Marcet's *Conversations in Chemistry*, fascinated young Faraday with the study and problems of chemistry, especially in exposing him to the idea of electrochemical effects requiring two electrical fluids, as shown in the important discoveries and theories of Sir Humphry Davy, and aroused his desire to make simple experiments and to speculate on the relation of chemical to electrical forces.

A great influence toward a more determined dedication to science was Faraday's opportunity to read the section of the *Encyclopedia Britannica* dealing with electricity, which he was at that time binding. This article inspired him to construct and experiment with electrostatic machines, within the limits of the few pence per week he could afford to spend. During the evenings, in the rear of the Riebau shop where he had assembled a small laboratory, he constructed a generator using electrostatic friction with a rotating bottle as the charge-generating surface, and another bottle to construct a Leyden jar.

From Watt's teachings, Faraday became aware of the fact that books alone were not adequate for the total education of an individual, and that human contacts with opportunities for discussion were also necessary. While employed as an apprentice, he noticed advertisements in shop windows publicizing a series of popular science lectures by a Mr. Tatem, to be held in the evening. The result of his attendance at a lecture in February 1810 was a desire to attend additional lectures. With the financial assistance of his brother, he was able to fulfill this desire and attend a series of twelve lectures. Through Mr. Tatem, the young

Faraday made the acquaintance of two intelligent men who had a great influence on his self-education. One was Mr. Huxtable, a medical student; the other Mr. Benjamin Abbot, a city employee. Huxtable loaned him two books, the third edition of Thomson's *Chemistry*, and Park's *Chemistry*; the latter Faraday in turn bound for his friend. Discussions and correspondence with Huxtable widely improved his own understanding of the modest experiments in which he had been engaged. Abbot maintained continuous correspondence and discussions with Faraday concerning Faraday's studies and work over an interval of many years. This friend served as a critical audience during his book-learning and experimental period.

Mr. Masquerier, a lodger in Mr. Riebau's house, loaned a copy of Taylor's *Perspective* to the young Faraday, who closely studied and copied all of the drawings. From the description and analysis, he made other geometric patterns, applying the stated rules in the exercises.

Faraday, having in mind Dr. Watt's teachings, followed the excellent practice of recording his thoughts, reflections and work in notebooks. In his earliest notebooks he recorded the names of the books and subjects which interested him most, together with his personal comments. He called these notes his "Philosophical Miscellany," being a collection of notices, occurrences and articles that were related to the arts and sciences. Faraday wrote of his desire to be able at some time to corroborate, or invalidate, the theories which were continuously startling the world of science and which were announced in various articles.

The City Philosophical Society which Mr. Tatem had organized in 1803 with a Mr. Magrath as secretary, had about thirty members and met every Wednesday evening for the purpose of mutual instruction. The fact that it was a society for mutual instruction was one of the outstanding aids in the pattern of the self-education of Faraday, who had joined the organization and had become an active member. To Faraday, Tatem served the valuable function, equivalent to a secondary school counselor, at a time when this guidance would be most effective.

For every second meeting, members and friends were invited to attend a lecture required of each member. The discussions and correspondence which resulted from these presentations were of invaluable assistance to Faraday in his development from tradesman to scientist.

Now came the most important event in Faraday's life; the fulfillment of an intense desire to leave trade and be engaged in scientific pursuits. This began when Mr. Dance, a customer of Riebau and a member of the Royal Institution, who was impressed with Faraday's quest for knowledge in the sciences, took him to four lectures by Sir Humphry Davy. Faraday absorbed all that he had heard and seen, and wrote a comprehensive set of notes on these lectures. It was of importance for Faraday to hear of and see experimental work of the outstanding scientist of the era. Faraday had read Davy's article on galvanism in the "Chemical Observer" which dealt with the electrolytic decomposition of alkalies, leading to the isolation and production of metallic potassium. Encouraged by Mr. Dance, he wrote Davy of his desire to be employed in the laboratories of the Institution, enclosing notes which he had written at the four lectures. Davy was apparently much impressed and agreed to meet with Faraday. In this interview, Davy told Faraday that although he was favorably impressed, no opening existed at that time, but if one should occur Faraday would be given consideration.

By this time Faraday had completed his apprenticeship and worked as a journeyman bookbinder for a Mr. DeLaRoche. He had grown to despise this trade, for now his dedicated goal in life was to be engaged in scientific work.

Three months after applying for employment at the Royal Institution, Faraday's much wished-for opportunity finally arrived. Davy's laboratory assistant had been discharged because of misconduct. Davy highly recommended to the board of managers that Faraday be engaged and they concurred with this proposal.

Of tremendous importance to Faraday was that this employment gave him contact with a source of continuous inspiration, and at the same time a mentor, during a formative period when the influence of this close association would be most effective. Davy had acquired an international reputation as one of the world's most outstanding chemical philosophers. He was a brilliant lecturer and the author of *Elements of Chemical Philosophy*. He had made a number of basic discoveries, as exemplified by his electrochemical isolation and production of metallic potassium and sodium in 1807. The new employment also provided Faraday with the opportunity to use the library of the Royal Institution and this was to be of inestimable value in his self-education.

The skills acquired from his experiments at home and the knowledge and experience derived from The City Philosophical Society enabled Faraday to be of immediate effective service when engaged by Davy as his chemical assistant. In addition to assisting Davy, Faraday also helped to set up the lecture demonstrations to be given by others and closely observed their presentations. He rapidly recognized the importance of the proper delivery of a lecture, together with the requirements beyond a mere knowledge of the subject.

Having an inherent drive for perfection in the preparation for his future activities, he formed with Magrath a mutual improvement group which met one evening each week in his room in the attic of the Royal Institution. Half a dozen or so men, chiefly from The City Philosophical Society, joined them to read together and to improve each other's pronunciation and construction of language. Although the discipline was severe, the results were so valuable that the group continued to meet for several years. Besides his strict self-discipline, one outstanding characteristic which Faraday possessed was his ingrained habit, fortified by the teachings of Dr. Watt, of maintaining continuous notes, with full disclosures and complete sketches, throughout his life. He had a keen sense of observation combined with imaginative thinking, and this was the root of an abundance of ideas governed by a sound judgment, reinforced by his unsparing self-criticism. His letters to his friend Abbot, whose good formal education Faraday respected, reflect the encouragement given him by Abbot. This correspondence served as an outlet to describe his experimentation, theorizing and analyses. In one of his early letters to Abbot, Faraday described his experiments in constructing electrical cells and batteries together with his observations of the electrolytic decomposition of salts of elements such as magnesium and copper. These letters gave him an opportunity to express his own concepts of his observations and to question the accepted nomenclature and theories of his time.

During his association with The City Philosophical Society, and two and a half years after he was employed by the Royal Institution, Faraday gave a series of eighteen lectures as a course in chemistry. His first six lectures indicated the wide range of his understanding, covering such subjects as the attraction of cohesion, chemical affinity, radiant matter, oxygen, chlorine, iodine, fluorine, hydrogen and nitrogen.

His lectures at The City Philosophical Society were related to physics as well as to chemistry of matter. In his fourth lecture on radiant matter he stated, "Assuming heat and similar subjects to be matter, we shall then have a very marked division of all the varieties of substance into two classes; one which will contain ponderable, and the other imponderable, matter. The great source of imponderable matter, and that which supplies all the varieties, is the sun, whose office it appears to be to shed these subtle principles over our system." His creed as a philosopher, stated in his fifth lecture, was that "a philosopher should be a man listening to every suggestion, but determined to judge for himself. He should not be biased by appearance, have no favorite hypotheses, be of no school, and in doctrine have no master. Truth should be his primary object."

Faraday was admired and respected by his fellow members of The City Philosophical Society. One of them, Mr. Dryden, published a poetic tribute to him in *The Quarterly Night* of October 2, 1816. It reads:

> But hark! a voice rises near the chair!
> Its liquid sounds glide smoothly through the air;

> The listening muse with rapture bend to view,
> The place of speaking and the speaker too,
> Neat was the youth in dress, in person plain;
> His eye read thus, philosopher in grain;
> Of understanding clear, reflection deep;
> Expert to apprehend and strong to keep.
> His watchful mind no subject can elude,
> No specious arts of sophists ere delude.
> His powers unshackled range from pole to pole;
> His mind error free, from guilt his soul.
> Warmth in his heart, good humor in his face,
> A friend to mirth, but foe to vile grimace;
> A temper candid, manner unassuming,
> Always correct, yet always unpresuming.
> Such was the youth, the chief of all the band;
> His name well known, Sir Humphry's right hand.
> With manly ease toward the chair he bends,
> With Watt's logic at his finger ends.

His education and development while in the laboratories of the Institution were very rapid. The opportunities of assisting at and of listening to the lectures presented at the Royal Institution by the authorities of that period were of immeasurable value. Of equally significant help was the encouragement and mentorship given to him by Davy. In his journal, Faraday stated that "in the glorious opportunity I enjoy of improving in the knowledge of chemistry and science with Sir Humphry Davy I have learned just enough to perceive my ignorance and the little knowledge I have gained makes me wish for more."

Of particular advantage to Faraday was the opportunity to join Davy on a trip through the Continent which lasted eighteen months. During this trip, he assisted Davy in the presentation of experimental work and in cooperative activities with many of the leading and prominent scientists of that era, including men such as Ampère, Volta, Chevreul, Humbolt, Gay-Lussac and especially, De la Rive. While in Paris, during their meeting with

Clement and Desmorne, they were shown a new elemental substance discovered only two years earlier by M. Courtois, a saltpeter manufacturer, and considered a complex form of chlorine. This substance became the subject of much investigation by Davy and later, in the form of organic compounds, by Faraday. Davy later identified it as Iodine.

The importance of this work and the further investigation can be noted in a letter Faraday sent to his friend, Huxtable.

> Sir Humphry Davy made his route as scientific as possible, and you must know that he has not been idle in experimental chemistry, and, still further his example did great things in urging the Parisian chemists to exertion. Since Sir H. has left England he has made a great addition to chemistry in his researches in the nature of Iodine. He first showed that it was a simple body. He combined it with chlorine and hydrogen, and latterly with oxygen and then added three acids of a new species to the science. He combined it with the metals and found a class of salts analogous to the hyperoxymuriates. He still further combined these substances, and investigated their curious and singular properties. The combination of iodine with oxygen is a late discovery, and this paper has not perhaps reached The Royal Society. This substance has many singular properties. It combines with acids and alkalies forming crystalline acid bodies and with the alkaline metals oxyiodes, analogous to the hyperoxymuriates. It is decomposed by heat about that of boiling oil into oxygen and iodine, and leaves no residue.
>
> It confirms all Sir H's. former opinion and statements, and shows the inaccuracy of the French chemists on the same subject.

During their stay in Paris, Davy and Faraday attended lectures given by Gay-Lussac at the L'École Polytechnique. At the home of a chemist in Genoa, the pair observed experiments with electric fish (torpedoes). This subject would later be of interest to them for their own experimental activities relating to the study of animal electricity. In Florence, Faraday was greatly excited at seeing the first telescope built by Galileo who had used it to discover the satellites of Jupiter. At the Academia del Cimento, by means of large optical lenses which they had borrowed, they investigated the combustion of diamonds. As a result of identifying the combustion product as pure carbon dioxide, the pair established that diamonds were composed of pure elemental carbon. The combustion of the diamond was an exciting experience for Faraday. He described the apparatus they used, comprised of a glass globe 22 cubic inches in volume which was exhausted of air and filled with pure oxygen, obtained from potassium chlorate. The diamond was supported in the center of this globe by a rod of platinum to the top of which a cradle or cup was affixed, perforated so as to allow free circulation of the oxygen about the diamond. The burning lenses composed of two double convex lenses, were at a distance from each other of about $3\frac{1}{2}$ ft. The large lense was about 15 inches in diameter, the smaller one about 3 inches in diameter. The diamond was placed in focus, which heated it intensely when the sun shone brightly. The amount of heat radiating from the heated diamond required cooling of the glass globe at times.

Sir Humphry Davy observed that the diamond was burning visibly and when removed from the focused solar heat continued in a state of active combustion. The diamond glowed brilliantly with a scarlet light inclining to purple, and when placed in the dark continued to burn for four minutes. The heating was repeated several times until the diamond was all consumed. The carbonic gas thus formed proved that the diamond is pure carbon.

While in Milan, Faraday had the wonderful experience of meeting Alessandro Volta, the inventor of voltaic cells and batteries, who had come to visit Davy. Volta's discovery of the means for chemically producing electrical energy, the Volta cell, was the prime source of electricity for the electrochemical experimentation of Davy and Faraday.

Faraday made a favorable impression on the scientists he met and maintained correspondence and an exchange of ideas with them throughout the remainder of his life, particularly with De la Rive.

There is another facet of Faraday which can be noted from the meticulous journal which he kept during his continental trip. In it can be recognized the latent poet, for his descriptions of the beauties of nature and the spectacular scenes he viewed especially in Switzerland, are outstanding in terms of his power of expression. In many cases, his descriptions are more impressive than paintings. These expressions illustrate how keenly observant and appreciative he was of beauty and of the magnitude of nature. This was later reflected in a philosophical lecture in which he said, "A sensitive mind will always acknowledge the pleasure it receives from a luxuriant prospect of nature. The beautiful mingling and graduation of colour, the delicate perspective, the ravishing effect of light and shade and the fascinating variety and grace of the outline must be seen to be felt, for expressions can never convey the ecstatic joy they give to the imagination or the benevolent feeling they create in the mind. There is no boundary, there is no restraint till reason draws the rein and then imagination retires into its recesses and delivers herself up to the guidance of that Superior Power." Faraday's analysis of the characteristics of the people in the areas he visited is an interesting comment on them and their society during this period in which, incidentally, he also learned to read French and Italian with facility. Faraday had the important opportunity of assisting Davy in the revision of his book by copying changes brought about during their discussions. This experience, besides being instructive, was invaluable when the time came for Faraday to publish his own books.

Faraday's return to London and continuation of his work as chemical assistant, together with his intense study habits, very rapidly increased his skill and knowledge to the status of expert. In addition to his duties as chemical assistant, he was also given supervision of the laboratory apparatus and of the mineralogical collection. He had a desire to become more fluent in presenting lectures and consequently, prior to taking part in Professor Brande's laboratory lectures at the Royal Institution, he took evening lessons in elocution from a Mr. Smart. These he continued for several years, up to the time when he presented his own first course in the theatre of the Institution. Mr. Smart often attended these lectures in order to provide Faraday with constructive remarks on his address and delivery.

In this profile I have referred to the imaginative thinking of Faraday as an outstanding characteristic. He recognized the importance of imagination and in an essay written in 1818, in which he referred to it as associated with judgment, he wrote that the function of imagination is to produce all kinds of possibilities and analogies and lay them before the mind. The senses provide the means for examination of the external world. Judgment is the key faculty. It is judgment which accepts the gift of imagination but then subjects it to the most intense critical scrutiny. The evidence provided by the senses allows judgment to control the imagination and extract from its offering what appears reliable and reject what cannot pass the acid test of experience. He considered that education is the process of training the judgment.

Faraday did not always accept prevalent theories as fact. For example, he did not believe white light was transmitted through accidental holes in a thin metal film. He explained this in his fourth lecture on radiant matter as follows. "The metals are among the most opaque bodies we are acquainted with, yet when beaten in very thin leaves they suffer light to pass. Gold, one of the heaviest of metals when beaten out and laid upon glass forms a screen of much transparency and anything strongly illuminated as by the sun may be seen through it. It has been said that this is occasioned by existence of small holes in the leaves, which permit the light to pass through them and that it does not pass through the body of the metal. If by small holes he meant the pores of the metal the explanation will readily be granted, but then the metal must be considered to a certain degree transparent for the transmission of light through the pores is the only way in which it can be transmitted at all; and nothing else takes place in transparent bodies; but if it be said that the existence of such holes as the light is supposed to pass through is accidental and only happens when the leaves are very thin, then arguments can be opposed to such a statement; for supposing it to be true, the light which passes should be white, whereas it is colored, and the color is found to depend on the metal being influenced by other substances which it may contain. Pure gold appears by transmitted light of a purple color; gold with a little silver, bluish, with a little copper, green, with iron, red; and these changes of color almost prove that light does not pass through such small accidental holes but actually through the pores of the metal, as with other metal.

In May, 1821, after Faraday was appointed superintendent of the house and laboratory of the Royal Institution, he was married, in the simplest of ceremonies, to Sarah Barnard, the daughter of an elder of the Sandamanian Church. The Faradays lived in an apartment at the Royal Institution.

After Faraday had acquired an excellent training by the combination of self-education and experience at the laboratory of the Royal Institution, he contributed many important discoveries, some of which have been basic factors in the industrial evolution. In line with the teachings of Watt, he believed experimental evidence was the most important basis for the truthful interpretation of scientific inquiry. He was rarely discouraged by failure to arrive at an anticipated result in an experiment and would repeat the procedures with such variations as he believed would correct the process. In most cases this resulted in an ultimately successful experiment. One can perceive the thread of continuity in all of his work on chemical, physical and electrical developments from his first experiments to his last, all so well preserved for posterity by the meticulously kept journals, diaries, papers and books.

After chemistry, his major interest was in magnetic phenomena and electricity. One of the experiments well known to physics students is the Faraday Cage, in which he proved that within an electrically conductive container there would be no electrical manifestation regardless of how intense an electrical charge was applied externally to such a unit. Faraday demonstrated the operation of induction from electric fields by carefully arranging an insulated metal chamber connected by a wire to a gold leaf electroscope. When an insulated but electrically charged brass ball was lowered into the cylinder the leaves of the electroscope diverged. After a certain depth the divergence remained constant. Upon withdrawal of the ball, the leaves of the electroscope collapsed.

Faraday applied the term "dielectrics" to insulating materials to express the fact that they allowed electrical forces through them as contrasted to conductors which will not allow electrical fields to penetrate their interior. He drew an analogy to the magnetic field shown by the displacement of iron fillings in a polar arrangement along the magnetic lines of force. The electrification of a dielectric represented to him a dielectric polarization. This understanding of the electric field effects on dielectrics enabled him to show that the theory prevalent at his time, based on action at a distance by the effect of a charged plate, was not true. He showed that capacitance in a condenser was due to the polarization of the dielectric materials in an electric field and he coined the term "specific inductive capacity" as a constant of dielectric materials. He did this by arranging two concentric spheres spaced and insulated from each other, and by evacuating the space he could introduce gases or liquids which would vary the capacitance according to the specific inductive capacity, or dielectric constant, of the spacing materials. In recognition of his basic research work on dielectric constants, the term "farad" was adopted as the unit of capacitance.

His work for the Electric Telegraph Company on the problem of transmitting signals through submerged insulated wires was of value to them. He demonstrated that the wires had a conductive wire enclosed by an insulator covering in contact with the sea water acting like a condenser, a charge developing in the insulation equivalent to that of a Leyden jar. A number of years later, the problem became a limitation to telephone transmission and it was another Michael, Michael Pupin, who solved the problem by insertance of inductance loading coils to change the phase relations introduced by the line capacitance.

The eminent scientist, James Clerk Maxwell, established support for Faraday's Field Action Theories when in 1855 he presented to the Cambridge Philosophical Society a paper entitled "On Faraday's Lines of Force." He acknowledged his debt to Faraday in a letter to him in which he stated that to his knowledge, Faraday was the first person to recognize the theories of bodies acting at a distance by throwing the separating medium into a state of constraint.

Faraday's work on magnetism led him to demonstrate the magnetic properties of oxygen in relation to other gases. At that time he demonstrated that oxygen, when heated, lost some of its magnetic characteristic of being influenced by a magnetic field, in the same way as iron or nickel. He showed that this temperature for oxygen was within the common temperature ranges of the atmosphere, and that this characteristic increased when cooled to 0°F or below. He believed that oxygen in the atmosphere was so magnetic in comparison with the other gases present that it was a probable source of the periodic variations in terrestrial magnetism. His work on the magnetic effects of oxygen-filled soap bubbles created great interest and enchanted his audiences. Faraday made the important discovery that all solids and liquids which he examined were either attracted or repelled by a powerful electromagnet. The bodies which were attracted were called "paramagnetic," and those repelled, "diamagnetic." The first diamagnetic effects were observed by Faraday in a particular glass called "Faraday's Heavy Leaded Glass."

Another dramatic experiment made by Faraday, to illustrate in his lectures the magnetic properties of gases, was the mixture of a small quantity of a visible gas or vapor with the gas under examination. The added visible gas or vapor rendered it perceptible when it ascended between the two poles of a magnet, and he observed their deflections from the vertical line in the axial or equatorial direction. In this way he found oxygen the least, nitrogen more, and hydrogen the most diamagnetic. With iodine vapor produced by placing a little iodine on a hot plate between the poles of the magnet the repulsion was strongly marked.

His interest in self-induction effects was increased when one of his assistants, M. W. Jenkins, mentioned to him his experience in getting an electric shock when holding both of the uninsulated wires of an electromagnet while making or breaking the circuit from a single voltaic cell. Since Jenkins had no intention of investigating this effect, Faraday decided to carry the subject further. He determined that the higher voltage pulsation obtained when an electromagnetic circuit was made or interrupted was caused by induction in its own circuitry, which could be in the same circuit or a secondary one electromagnetically coupled to it. He had observed the bright sparks that were generated when an electromagnet circuit was interrupted, and was now able to identify its cause as self-induction. He noted that the rise in potential and change of direction of current flow in a secondary circuit occurred only when the primary circuit was made or interrupted.

After Faraday had studied his discoveries in the production of electric currents by induction from a magnetic field the significance of the work reported by Arago became clear to him. Arago had cited the dampening effect he noted on the oscillations of the magnetic needle in a compass when it was placed on a copper disc, and the movement of the magnetic needle if the compass were placed near a revolving copper disc. To Faraday, this unexplained phenomenon meant that induced currents in the copper disc produced electromagnetic fields that opposed the relative motion of the magnetic needle. He was confident that a magnetic force would affect a beam of light, and after twenty years of experimentation he found that when he placed a heavy piece of lead glass in a magnetic field, and when the magnetic lines of force were parallel to the direction of a beam of polarized light, the plane of polarization rotated. His work on magnetic field effects on light was basic for the later developments by others in the field of magneto-spectroscopy. His studies of diamagnetic and paramagnetic materials, particularly with the crystallinity of bismuth in relation to its diamagnetic properties, along with studies of the differential magnetocrystallic forces in different media and the reaction to heat, were of value to the understanding of the science of materials.

In 1821 he discovered the principle of the electric motor by causing a wire, through which current was discharged, to rotate around a magnet.

Acting upon Davy's recommendation of February 7, 1825, the Board of Managers of the Royal Institution appointed Faraday Director of the Laboratory. The effect of this well deserved recognition was to give Faraday greater freedom to carry out his own concepts for researches in electricity. Many universities sought Faraday's services as a professor of chemistry, but he maintained his loyalty and attachment to the Royal Institution.

The most important example of Faraday's persistence of effort in translating his imaginative thinking into a reality was his conversion of magnetic forces by induction into electrical energy. He persistently held the opinion that since electricity could produce magnetism, the reverse should be possible. In the fall of 1831 he succeeded in producing an electric current flow in a conductor by induction from a magnetic field. The first work was with an electromagnetic field by discharging a battery through a coil and later, by moving a bar-type permanent magnet into a helix. He had found that the essential requirement for conversion of magnetism into electricity was that a relative motion or change in the density of the magnetic field could, by induction, cause the generation of electric currents.

He then rotated a copper disc between the poles of a horseshoe magnet and connected the edge of the disc and the axis to a galvanometer and produced an electric current while the disc was rotated. He also confirmed the basic factor that it required a variation in the relation of a magnetic field to a conductor in order to induce electric current in the inductively coupled conductor, this being accomplished by the interruption of current through an electromagnet or the movement of a conductor across the magnetic lines of force of a magnet. He also found that by reversing the polarity of the magnetic field, he obtained a reversal of the current induced, as indicated by his galvanometer, the same effect occurring if the copper disc were rotated in the opposite direction. Faraday discovered that terrestrial magnetism can develop induced currents in conductors in motion. He rotated a coil in the earth's magnetic field and noted the alternating electric force generated. By rotating a rectangular coil in the earth's magnetic field and providing a commutator to its terminals, he was able to obtain a unidirectional current. All these developments were the forerunners of the commercial dynamo of industry developed some thirty-five years later. No achievement in the history of technology has had a greater impact on civilization than did this means of generating electric currents by rotating conductors in a magnetic field, or the induction of current in one coil from another by variation of the exciting electromagnetic field. Sir Robert Peel, then Prime Minister, visited Faraday in his laboratory at the Royal Institution and seeing a model of the magneto-electric generator inquired, "of what use is it?" Faraday is said to have replied, "I know not, but I wager some day your government will tax it." How true this prophesy was, for with the commercialization of magneto-electric-produced electricity since 1880, the government income from electric utility taxes has become an increasing source of revenue.

It was the magneto-electric generator that made industrial electrochemistry possible. Since industrial applications required high current densities that could not practically be obtained from batteries, the development of direct current generators afforded a practical source of power for large-scale electrochemical operations.

In 1831 he wrote the first section of his *Experimental Researches in Electricity*, that would occupy him for the next 23 years. At first this work was published as monographs in the

Transactions of the Royal Society, and later published as three separate volumes in 1844, 1847 and 1855. Since publication, this work has served internationally as an inspiring source of basic studies in electrical research. His studies of the conductivity of crystals are of interest to those engaged today in solid state physics. His books reflect his genius in applying imaginative use and coordination of facts of science stored in his creative mind. This extraordinary work has served as a source of inspiration and a guide to those who followed him in experimental work in electricity and chemistry over the years.

Faraday conveyed part of his philosophy of motivation when in a lecture he stated, ". . . in the pursuit of science we first start with hope and expectations. These we realize and establish, never again to be lost, and upon them we found new expectations of further discoveries, and so go on pursuing, realizing, establishing, and founding new hopes again and again."

Faraday's creative work in physics, chemistry, and particularly electrochemistry, had great impact on the growth of scientific knowledge. His study involving the liquification of several gases, including chlorine and the diffusion of gases through solids, was of importance in the identification and knowledge of their properties. Faraday, working with a liquid condensate derived from The Portable Gas Company's oil gas, discovered benzene. In a continuation of this work, he produced nitrobenzene, o, p-dichlorobenzene, hexachlorobenzene, hexachloroethane, naphthalene, monosulfonic acid, butylene, and chlorine derivatives of butylene, and many other halogenative organic compounds. Working with Stodart, surgical instrument maker, he established the requirements for producing nickel iron alloys. He found that it required iron with a low carbon content to alloy with nickel, whereas high carbon steel hindered the formation of a desirable alloy. The range of nickel content and carbon limitations were used in later years in the development of stainless steel.

Faraday held that chemical and electrical forces were identical. In 1834 he wrote in his notebook that chemical affinity is electricity. In his lecture at the Royal Institution on June 12, 1834, on the relation of chemical affinity, electricity, heat, magnetism, and other powers of matter he demonstrated that all were inter-related.

On July 28, 1836, under the subject of thermo-electricity he stated, "surely the converse of thermoelectricity ought to be obtained experimentally by passing a current through a circuit of antimony and bismuth."

One of Faraday's persistent interests was in proving that electricity from all sources had essentially the same action. He showed that electricity generated electrostatically by a friction machine would cause electrolytic decomposition just as would electricity generated by chemical action in voltaic cells or by magnetic induction in a moving conductor. In his study of electrolytic conduction, he investigated the electrical properties of such solid substances as chlorides, fluorides, sulfates, nitrates and others that were essentially non-conductive until molten or liquified, when decomposition by the passage of current took place. His observations on the anodic oxidation of copper electrodes was information that would be followed by others in the studies of voltaic cells. He found that such solid crystalline water-insoluble compounds as silver sulfide were slightly conductive, but when heated became increasingly conductive, with values approaching those of metals. Silver sulfide filaments have been used in modern devices to indicate temperature by change in electrical resistance.

Perhaps Faraday's most important contribution to the science of electrochemistry was his placing it on a quantitative basis culminating in his discovery of the basic relation of chemical change with current and time in electrolysis. In order to describe his research in terms that were more specific, he initiated descriptive terms such as "electrodes" instead of poles, "ions" as the conductive components in conductive liquids, "electrolytes," and the action of electrical decomposition, "electrolysis." He called the positive electrode the "anode" and the negative one the "cathode," the ions flowing to them being "anions" and "cations" respectively.

This latter group of terms was adopted after an exchange of ideas by correspondence with his friend Reverend H. Whewell Master of Trinity College, Cambridge University, who thought that they were more suitable than the philogical Greek derivations Faraday had at first planned to use.

He discovered that the same amount of current passed through a series of cells having different electrode sizes, and with different amounts of acid in the electrolytes, produced the same electrolytic product of decomposition. He found that the amount of material released from the electrolyte by electrolysis with a given current and time was dependent upon the chemical equivalent of the material.

Faraday needed better means for measuring current flow and electromotive force. He developed the voltameter, which measured the quantity of gas or solids liberated by an electric current in a measured time.

In one device he utilized the measurement of oxyhydrogen gas developed in his apparatus of 0.1741 cc. per second liberated at the electrodes by passage of one ampere second of current.

With solid electrolytic depositions he indicated .001181 grams of silver deposited per ampere second from a silver salt solution.

Faraday was one of the first scientists to place electrochemistry on a scientific and quantitative basis, and his discoveries are embodied in Faraday's Law of Electrolysis. In recognition of his work, the term "faraday" (F) was adopted, which by definition is the amount of electricity that will liberate a gram equivalent weight of a material released or deposited at the electrodes in an electrolyte. For practical purposes, this faraday unit is taken as 96,500 int. coulombs, or 96,500 ampere seconds, and the electrochemical equivalent in turn is the gram weight liberated by one coulomb. The discovery of the fundamental laws of electrochemistry had a far-reaching effect on the evolution of chemical technology and the development of the electrical theory of matter.

Faraday's researches established quantitatively the relation between electricity and atomic weight of the elements. He stated, "For a constant quantity of electricity, whatever the decomposing conductor may be, whether water, saline solutions, acids, fused bodies or the like, the amount of electrochemical action is a constant quantity. . . . with different solutions the amount of elementary constituents produced by unit quantity of electricity are proportional to their chemical equivalents."

Element	Atomic Mass	Valence	Chemical Equivalent	Electrochemical Equivalents Micro grams per Coulomb	Equivalents Coulomb per gram
Hydrogen	1	1	1	10.38	90,340
Oxygen	15.96	2	7.98	82.83	12,070
Chlorine	35.37	1	35.37	367.10	2,724
Nitrogen	14.01	3	4.67	48.47	20,630
Aluminum	27.04	3	9.01	93.50	10,700
Lead	206.40	2	103.2	1,071	933.7
Zinc	64.88	2	32.44	333.70	2,970
Nickel	58.60	2	39.30	304.20	3,287
Mercury	199.80	2	99.90	1,037	964.3
Mercury	199.80	1	199.80	2,074	482.2
Copper	63.18	2	31.50	327.90	3,050
Copper	63.18	1	63.18	655.80	1,525
Silver	107.70	1	107.70	1,118	894.5
Gold	196.2	3	65.40	678.90	1,473

Table showing the relative weights of material liberated by the flow of an electric current as found according to Faraday's discovery.

His research on the conduction of salt solutions were the first experiments indicative of discrete units of electricity. Von Helmholtz, in his Faraday lecture of April 5, 1881 at the Royal Institution, corroborated this point of view when he gave a complete explanation of the facts embraced by Faraday's Law. He stated that "Every single valency of an elementary or compound ion is charged with exactly the same quantity of positive or negative electricity, which behaves as if it were an electrical atom that cannot be further divided."

"The same definite quantity of either positive or negative electricity moves with each monovalent ion, or with every unit of affinity of a multivalent ion and accompanies it during all its motions through the interior of the electric fluid. This quantity we may call the charge of the atom."

"If we accept the hypothesis that the elementary substances are composed of atoms we cannot avoid concluding that electricity also, positive as well as negative, is divided into definite elementary portions, which behave like atoms of electricity. As long as it moves about in the electrolyte fluid, each ion remains united with its electrical equivalent or equivalents. At the surface of the electrodes decomposition can take place if there is a sufficient electromotive force, and the ions give off their electrical charges and become neutral atoms."

The important table of electrochemical equivalents was based on Faraday's work. The determination of the electrochemical equivalent can be derived from the following formulae: Electrochemical equivalent $= \frac{kA}{n}$ in grams per amp. hr. where

k is the proportionality factor with a value of 0.0373100,
A is the gram atomic weight,
n is the valence

In Faraday's Law the atoms and electricity are intimately associated. Faraday's ions transferred through the solution in the process of conduction were carried in definite amounts, one gram ion for every 96,500 coulombs of electricity. It was modified by Grove in 1845, by Williamson in 1851 and by Clausus in 1857, in the sense that the decomposition was not affected by the current, but that the ion carriers must exist in part uncombined. The theory of conduction was given formulation by Arrhenius in 1887, who stated that the extent to which free ions occurred in solution was deducible from the electrical conductance of the solution. The properties of conducting solutions became a property of the positive and negative charges they contain. In the development of the electron theory of matter, it was G. Johnstone Stoney who, in 1891, gave the name "electron" to the natural unit of electricity defined in reference to Faraday's Law. Faraday's Law was not sufficient to establish the atomistic concept of electricity and it required the proof obtained by J. J. Thomson who, in 1897 at the Royal Institution, announced the discovery of the electron. This work had been accomplished at the Cavendish Laboratory in Cambridge, England, by ionization of gases with x-ray or radioactive materials.

To prove that electricity from a voltaic cell results from chemical action and not simply contact potential, Faraday applied a doctrine of the conservation of energy in 1840 by stating, "The contact theory assumes force which is able to overcome a powerful resistance can arise out of nothing. We have many processes by which the form of powers so change that an apparent conversion of one into the other takes place. But in no case is there a pure creation of production or creation of power without a corresponding exhaustion of something supplying it."

Our present awareness of cumulative pollution of our natural resources might be related to Faraday's protest on this matter. On June 7, 1855, he wrote a letter to "The Times" complaining about the pollution he observed in the river Thames.

Royal Institution, July 7, 1855
Sir—I traversed this day by steamboat the space between London and Hungerford Bridge, between half past one and two o'clock. It was low water and I think the tide must have

been near the turn. The appearance and smell of the water forced themselves at once to my attention. The whole of the river was an opaque pale brown fluid. In order to test the degree of opacity, I tore up some small white cards into pieces and then moistened them, so as to make them sink easily below the surface and then dropped some of these pieces into the water at every pier the boat came to. Before they had sunk an inch below the surface, they were undistinguishable, though the sun shone brightly at the time and when the pieces fell edgewise the lower part was hidden from sight before the upper part was under water. This happened at St. Paul's Wharf, Blackfriars Bridge, Temple Wharf, Southwark Bridge and Hungerford, and I have no doubt that it would have occurred further up and down the river. Near the bridges the feculence rolled up in clouds so dense, that they were visible at the surface even in water of this kind.

The smell was very bad, and common to the whole of the water. It was the same as that which now comes up from the gully holes in the streets. The whole river was for the time, a real sewer. Having just returned from the country air, I was perhaps more affected by it than others; but I do not think that I could have gone to Lambeth or Chelsea, and I was glad to enter the streets for an atmosphere which, except for the sink-holes I found much sweeter than on the river.

I have thought it a duty to record these facts, that they may be brought to the attention of those who exercise power, or have the responsibility in relation to the condition of our river. There is nothing figurative in the words I have employed, or any approach to exaggeration. They are the simple truth.

If there be sufficient authority to remove a putrescent pond from the neighborhood of a few simple dwellings, surely the river which flows so many miles through London ought not to be allowed to become a fermenting sewer. The condition in which I saw the Thames may perhaps be considered as exceptional, but it ought to be an impossible state; instead of which, I fear it is rapidly becoming the general condition. If we neglect this subject, we cannot expect to do so with impunity; nor ought we be surprised if, ere many years are over, a season gives us sad proof of the folly of our carelessness.

I am, Sir, your obedient servant,

M. Faraday

The dire results that Faraday predicted did indeed happen, for up to twelve years ago all fish and marine life had ceased to exist in the Thames, particularly in the 92-mile industrial stretch which stunk and was considered poisonous. Today, after an expenditure of 45 million dollars, the 210-mile Thames, including the 92 mile industrial stretch, is acknowledged to be the cleanest industrial river in Europe, with a high standard of purity. Fish now thrive along the entire length of the river. The water is constantly sampled, and a close vigil is kept at all discharge points. Investigators from countries all over the world which have serious river pollution problems found that under the authority granted by the Acts of Parliament all discharges of industrial wastes were licensed and a rigid standard of purity was soundly enforced. All sewage treatment and other polluting effluents were prevented from draining their wastes indiscriminately.

Along with the many consulting services to his government, which included giving lectures to cadets at the Royal Academy at Woolwich and for 30 years serving as adviser to Trinity House on the supervision of lighthouses, Faraday assisted in the introduction of a magneto electric light for the lighthouse beacon at Dungeness.

His celebrated chemical history of the candle was one of the series of Christmastime lectures he initiated at the Royal Institution. Giving these lectures to the young people was a great satisfaction to him over the years, and they were much appreciated by his audience.

An ecological cartoon of 1855: Faraday investigating the polluted Thames.

Both the Royal Society and the Royal Institution had tried in vain to have him accept the presidency of these organizations; however, he recognized his physical limitations and desire to utilize his time and energy in the manner best suited to his philosophy of work and time.

The year 1855 brought the series of experimental researches of electricity to a close. It had begun in 1831 with his greatest discovery, the induction of electricity and the evolution of electricity from magnetism. Other discoveries and researches that followed were the investigation of terrestrial magneto electric induction, equivalent end results from different sources of electricity, thermo-conducting powers of materials, thermo-electric properties, electrochemical decomposition, voltaic cells, self induction effects, nature of electric forces from electric fish (gymnotus frictionae), electricity from expanding gases or vapors, magnetization of polarized light, illumination of magnetic lines of force, magnecrystalographic relations of bismuth and other crystals, and comparative studies of magnetic field effects.

In 1856, when he was 64 years old, Faraday complained of his inability to do much writing and developed a forgetfulness which limited his activities. This was a recurrence of disability from overwork which had required him to rest fifteen years previously. In October of 1861 he wrote to the managers of the institution:

"It is with deepest feeling that I address you. I entered the Royal Institution in March 1813, nearly forty-nine years ago, and with exception of a comparative short period during which I was abroad on the continent with Sir H. Davy, have been with you ever since. During that time, I have been most happy in your kindness, and in fostering care which the Royal Institution has bestowed upon me. Thank God, first for all his gifts. I have next to thank you and your predecessors for the unswerving encouragement and support which you have given me during that period. My life has been a happy one, and all I desired. During its progress, I have tried to make a fitting return for it to the Royal Institution and through it to science. But the progress of years (now amounting in number to three score and ten) have brought forth first the period of development, and then that of maturity, have ultimately produced for me a period of gentle decay. This has been taking place in such a manner as to make the evening of life a blessing; for whilst increasing physical weakness occurs, a full share of health free of pain is granted with it; and whilst memory and certain other faculties of the mind diminish, my good spirits and cheerfulness do not diminish with them.

"Still, I am not able to do as I have done. I am not competent to perform as I wish the delightful duty of teaching in the Theatre of the Royal Institution, and I now ask you (in consideration for me) to accept my resignation of the Juvenile Lectures. Being unwilling to give up what has always been so kindly received and so pleasant to myself, I have tried the faculties essential for their delivery, and I know that I ought to retreat; for the attempt to relate (in those trials) the necessary points, brings with it weariness, giddiness, fear of failure and the full conviction that it is time to retire. I desire therefore, to lay down this duty. I may truly say that such has been the pleasure of the occupation to me, that my regret must be greater than yours need or can be.

And this reminds me that I ought to place in your hands the whole of my occupation. It is no doubt true that the Juvenile Lectures, not being included in my engagement as professor were when delivered by me undertaken as an extra duty, and remunerated by an extra payment. The duty of research, superintendence of the house and of other services still remains, but I may well believe that the natural changes which incapacitate me from lecturing, may also make me unfit for some of these. In such respects, however, I will leave you to judge and to say whether it is your wish that I shall remain as part of the Royal Institution. I am Gentlemen, with all my heart your faithful and devoted servant.

M. Faraday"

1862 was the last year of experimental research. On June 26 he gave his last discourse for the Royal Institution. He continued some experiments and lecture work, but felt too exhausted to do research work to the standard he had previously maintained. He had given lectures for 38 years, which was the life of the Royal Institution. He was an inspiring educator, as well as a very popular lecturer. In 1861 he resigned from his activities at the Royal Institution and with his wife moved from their quarters there to a house on the Green near Hampton Court, placed at his disposal by Queen Victoria.

A review of his life and accomplishments shows how very important the encouragement of imaginative thinking is, and how help to a creative-minded individual who has that inherent sense of imaginative thought can be of positive and enduring effect during his life. Wisdom was a basic characteristic in the life of Faraday in the proper utilization of his brain-stored information. His imagination of scientific second thoughts, like a seer, led him to anticipate results which later proved true.

In 1848 Faraday gave a course of seven lectures on the allied phenomena of the chemical and electric force. In one of the lectures relating to voltaic batteries, he expressed his imaginative mind when he stated "Here consider how Davy wedded to this power and triumphed over the alkalies in this room and of the combustion of zinc. It gives one a strange sensation as to what may be going on in a gas flame or a fire and hope that someday we may transfer their light and power to a distance, and use them at pleasure, laying on not gas, but the power of gas or oil and so having a lamp more wonderful than Aladdin's."

Our present electric light and its energizing power come from a distant power station utilizing the power of gas or oil. When this is thought about, it fulfilled a prophetic vision of Faraday.

He loved truth beyond all things and devoted his efforts to searching for it. His great inherent energy and drive were consistently applied to the translation of his imaginative thoughts into practical reality, resulting in many successes, and yet he remained humble, kind, and always considerate of others.

While he received many honors during his life—about 95 honorary titles and marks of merit—he was proudest of the title he felt he earned in 1824 when 33 years old, "Fellow of the Royal

Society." His election to the status of F. R. S. involved some sad moments for him, for when he was being considered by the committee for this honor he was opposed by Sir Humphry Davy. It has been reported that Davy was jealous of the attention directed to Faraday, but he was unsuccessful in his opposition. Faraday, being of magnanimous character, recognized the pettiness of Davy's reaction and held no bitterness toward him, but remained always grateful for the help Davy had given him in his rise from tradesman to scientist.

In a letter to Faraday in 1858 M. De la Rive asked him whether it was true that he was inspired with his first taste for chemistry by the reading of Mrs. Marcet's *Conversation in Chemistry* and whether this determined the course of his work. Faraday stated in reply that it was Mrs. Marcet's *Conversations in Chemistry* that gave him his introduction and foundation in that science. He added, he "could trust a fact but always cross-examined an assertion. So when I questioned Mrs. Marcet's book by such experiments I could find means to perform and found it true to the facts as I could understand them, I felt that I had got a hold of an anchor in chemical knowledge and clung fast to it."

During his life he had interests other than science and delighted in reading aloud to his wife and niece works of such authors as Byron, Macauley, Shakespeare, and Scott. He enjoyed going to the theatre and opera, and was fond of good music. Before his marriage he learned to play the flute and was reported to have memorized 100 songs. He collected many specimens of minerals and plants, observed the biologies of a variety of small animal life and maintained a froggery to supply frogs used in his work.

In his last lecture certain phrases illustrate the philosophical thinking of Faraday, such as, "We are placed by our Creator in a certain state of things resulting from the pre-existence of society combined with our own laws of nature. Here we commence our existence, our early career. The extent before us is long and he who reaches furthest in his time has best done his duty and has most honour; the goal before us is perfection, always within sight but too far distant to be reached. Like a point in the utmost verge of perspective it seems to recede before us and we find as we advance that the distance far surpasses our conception of it. Still, however, we are not deceived; each step we move repays abundantly the exertion made and more eager our own race the more novelties and pleasures we obtain.

"Some there are who on the plain of human life content themselves with that which their predecessors put in their possession and they remain idle and inactive on the spot where nature dropped them. Others exist who will enjoy the advantages in advance but are too idle to exert themselves for their possessions and they are well prescribed for the envy which their very sensibility and sentient powers engender within them at the sight of success of others. A third set are able and willing to advance in knowledge but they must be led and but few obtain the distinguished honor of being first in their plans and of taking lead of their generation, of their age, and of the world."

He died in August of 1867 at the age of 75, exhausted but calm in spirit. The entire world can be thankful that there has been a human being like Michael Faraday who never desired nor acquired monetary riches from his work, but rather untold spiritual riches in the sense of accomplishment in baring the truths of nature.

PART VI

Reflections

Both Humphry Davy and Michael Faraday fulfilled the hopes that were engendered when Count Rumford founded the Royal Institution in London. Their work contributed more than scientific discoveries which made the Institute internationally famous and laid the basis for the founding of quantitative electrochemistry. Their exceptional ability to translate the sciences in a brilliant and fascinating manner as lecturers helped to sustain the Institute. This was most important since its founder retired from the institute only a few years after its establishment and before the announcements made possible by the great discoveries that followed.

Our present electrically oriented civilization had its basic founding when Michael Faraday of the Royal Institution discovered the means for generating electricity by the induction effect in a conductive element from a magnetic field. It is for this reason that Count Rumford was included in a study of the founding of electrochemical science, since the activities at the Institution were carried out on the basis of his endowed philosophy.

The inspiration derived from Michael Faraday carried on through the years for such practical inventors as Thomas Edison, who gave credit to Faraday's book on experimental electricity as an inspiring source of information and education.

The writer, an inventor in the electrochemical field who over the past fifty years has collected a small library of the complete Faraday Diary, his books, first copies of his papers to the Royal Society, and handwritten letters, also feels indebted to the teachings and spirit of Michael Faraday, reflected in his life and accomplishments. His books on *Experimental Researches in*

Electricity and Chemical Manipulation are an inspiration to read and study. They contain a large source of information which provides a stimulus for ideas and understanding. The writer considers these books one of the most important sources for building a storehouse of facts that could be usefully integrated with other information to supply a retrieval base for the generation of new concepts and for synthesizing a specialized technology for the translation of imaginative concepts into realities. Great theoretical scientists acknowledge his introduction of the field theory, mathematized by James Clerk Maxwell. Among them was Albert Einstein, who kept a portrait of Faraday in his office.

Another concept presented in an early lecture must have caused reflection by later physicists when Faraday stated in respect to the notion of transmutation, once considered absurd, "Let no-one start at the difficult task and think the means far beyond him. Everything may be gained by energy and perseverance. Let us but look to the means which has given us these bodies and their gradual development; we shall thereby gain confidence to hope for a new and effective power for their removal from an elementary state."

Faraday also imagined that there might be a great universal principle from which gravity, heat, light, electricity, magnetism, or even life might come. He thought that gravity, like the localized attractions of chemical affinity, might be an electric field effect.

On Mar. 19, 1849 he wrote in the laboratory book, "Gravity, surely this force must be capable of experimental relation to electricity, magnetism and other forces so as to bind it up with them in a reciprocal and equivalent effect." He continued for many pages for examination and thought and ended the study by stating, "All this is a dream. Nothing is too wonderful to be true if it is consistent with the laws of Nature, and in such things as these, experiment is the best test of such consistency."

BIBLIOGRAPHY

Antoine L. Lavoisier. *Traité Élémentaire de Chimie.* Paris: Imprimerie Imperiale, 1789.

Humphry Davy. *Researches Chemical and Philosophical Concerning Nitrous Oxide.* Bristol: J. Johnson & Co., 1800.

Isaac Watts. *The Improvement of the Mind.* London, 1809.

Jane Mercet. *Conversations in Chemistry.* London, 1809.

Humphry Davy. *Elements of Chemical Chemistry.* London: Johnson & Co., 1812.

Humphry Davy. *Elements of Agricultural Chemistry.* London: Longmans & Co., 1813.

R. Hunter. *Humphry Davy on the Safety Lamp for Coal Mines with Some Researches on Flame,* 1818.

Humphry Davy. *Six Discourses Delivered Before the Royal Society.* London: Murray, 1829.

Michael Faraday. *Chemical Manipulations.* London: Murray, 1829.

J. A. Paris. *The Life of Sir Humphry Davy.* London: Colburn & Bentley, 1831.

John Davy. *Memoirs of the Life of Sir Humphry Davy.* London: Longmans, 1836.

Humphry Davy. *Collected Works.* Ed. John Davy. London: Murray, 1839.

Michael Faraday. *Experimental Researches in Electricity.* London: Taylor and Francis, 1855 (American edition in two volumes, New York, Dover Publications).

Michael Faraday. *Experimental Researches in Chemistry.* London, 1859.

John Tyndall. *Faraday As a Discoverer.* London: Longmans Green, 1868.

Bence Jones. *Life and Letters of Faraday.* London: Longmans, 1870.

Bence Jones. *The Royal Institution, Its Founder and Its First Professors.* London, 1871.

George Ellis. *Life of Count Rumford.* American Academy of Arts and Science, 1871.

T. E. Thorpe. *Humphry Davy, Poet and Philosopher.* London: Cassell, 1896.

Atkinson and Reinhold. *Ganot's Physics,* 16th ed. London: Longmans, Green & Co., 1902.

Michael Faraday. *Faraday's Diary* (released by the Royal Institution). London: G. Bell & Sons, 1932.

Victor Robinson. *Victory Over Pain.* London: Sherman, 1946.

G. A. Foote. *Sir Humphry Davy and His Audiences at the Royal Institution.* London: ISIS, 43 no. 1, June 12, 1952.

Luigi Galvani. *Commentary on the Effects of Electricity on Muscular Motion.* Norwalk, Conn.: Burndy Library, 1953.

Sanborn C. Brown. "Count Rumford." *American Science,* June 1954.

J. F. Kendall. *Humphry Davy, Pilot of Penzance.* London: Faber, 1954.

Schwartz and Bishop, ed. *Moments of Discovery.* Vol. 2. New York: Basic Books, 1958.

W. J. Sparrow. *Count Rumford of Woburn, Mass.* New York: Crowell Co., 1964.

Bern Dibner. *Alessandro Volta and the Electric Battery.* New York: Franklin Watts, 1964.

L. Pearce Williams. *Michael Faraday.* New York: Basic Books, 1970.

Harold Hartley. *Studies in the History of Chemistry.* Oxford: Clarendon Press, 1971.

Bern Dibner. *Luigi Galvani.* Norwalk, Conn.: Burndy Library, 1971.

Stephan Longstreet. *We Went to Paris.* New York: Macmillan Co., 1972.

Rumford Historical Association. *Count Rumford, Sir Benjamin Thompson.* Woburn, Mass., 1972.

INDEX

A

Abbot, Benjamin, 68, 70, 71
Academia del Cimento, 74
Alkali, 54
Alkaline earths, 54
Alsace, 17
American Academy of Arts and Sciences, 21
Ampère, Andre Marie (1775-1836), 57, 72
Ampère's Rule, 59
Anesthesia, 49
Animal electricity, 30
Anions, 84
Annalen der Chemie, 20
Anodes, 84
Appleton, John, 13
Apreece, Jan, 55
Arago, Francois Jean (1786-1853), 80
Arc light, 59
Arrhenius, 88
Auteuil, 24

B

Bakerian Lecture, 53, 59
Baldwin, Loammi, 14
Ballistics, 16
Banks, Sir Joseph, 21, 22, 39, 51
Barnard, Sarah, 77
Baronetcy, 58
Bartells, Prof., 39
Bavaria, 17
Beccaria, Giovani Batiste, (1716-1781), 38
Beddoes, Dr., 46, 48
Benzene, 83
Bernard, Reverend Thomas, 14
Berthollet, 57
Berzelius, Jon Jakob, (1766-1852), 54
Bible, 66
Bismuth, 84
Bologna Academy of Science, 30, 32
Bonaparte, Napoleon, 23, 39, 41, 54
Borlase, 46, 48
Boscovitch, R. J., (1711-1787), 52
British Archives, 16
Burndy Library, 4, 29
Butylene, 83
Byron, 95

C

Caloric Theory, 19
Cambridge Philosophical Society, 63
Cambridge University, 88
Candlepower, 24
Capen, Hopestall, 14
Carbon, 48, 74
Carlisle, 37
Carlyle, 49
Cathode, 84
Cations, 84
Cavendish, 37
Cavendish Laboratory, 88
Chemical Balance, 46
Chemical Observer, 69
Chevreul, 57, 72
Chlorine, 83
City Philosophical Society, 68, 70, 71

Clausus, 88
Clement, 72
Coleridge, 45
Committee of Correspondence, 15
Como, 35, 38, 41
Compton, Arthur, 21
Compton, Karl, 21
Concord, 15
Condensers, 27
Condensing Electrophorus, 28
Condensing Electroscope, 28
Copley Medal, 51
Copper disc, 80
Coulomb, 85
Count of the Holy Empire, 19
Courtois, 73
Cruikshank, Dr. William, 37, 50
Crystals, 83

D

Dalton, John, 52
Dance, 57, 69,
Davy, Lady, 56, 57
Davy, Sir Humphry (1778-1829), 22, 29, 33, 37
De la Rive, August Arthur (1801-1873), 45, 72, 75
De la Rive, Charles Gasparde (1771-1834), 45
De LaRoche, 69
Dematoritus, 53
De Morveau, Guyton, 45
Descartes, Rene (1596-1650), 53
Desmorne, 72
Diamagnetic, 81
Diamonds, 59, 74
Dichlorobenzene o, p, 83
Dielectric, 78
Dielectric constant, 78
Dragoons, Royal, 16
Dryden, 71

Duke of Bavaria, 19
Dulong, 56
Dungeness, 90
Dunkin, Robert, 47
Dynamo, ε

E

École Polytechnique, 73
Edison, Thomas A. 21, 99
Einstein, Albert, (1879-1955), 100
Electric generators, 81
Elector, Karl Theodore, 17
Electric Telegraph Company, 78
Electrode, 84
Electro-deposition, 37
Electrolysis, 84
Electron, 88
Electroscope, gold leaf, 76
Electrostatic machines, 27
Elements of Chemical Philosophy, 24
Encyclopedia Britannica, 67
English Gardens, 18
Experimental Researches in Electricity, 82

F

Fabroni, 38
Farad, 78
Faraday Cage, 77
Faraday's Law, 85, 87
Faraday, Michael (1791-1867), 3, 57
Fermi, Enrico, 21
Field Action Theory, 79
Florence, 41, 74
French Military Academy, 23
French National Institute, 23
Friction, 27
Frog's leg reaction, 30
F.R.S., 16, 95

G

Gage, General, 15
Galilei, Galileo (1564-1642), 74
Galvani, Luigi (1737-1798), 30, 32, 33, 34
Galvanism, 33
Gay-Lussac, Joseph (1778-1850), 57, 72
Geneva, 60
Genoa, 73
Germaine, Lord George, 15, 16
Gibbs, Josiah Willard, 21
Giddy, Davies, 22, 46, 48
Gilbert, 46
Glass, heavy leaded, 79
Gold leaf light transmission, 76
Gram ion, 87
Grove, 88
Gymnotus Frictionae, 92

H

Harvard College, 14
Harvard University, 24
Hay, Dr. John, 14
Helices, 59
Helmholtz, Herman Ludwig von (1821-1894), 87
Herculeum, 58
Hexachlorobenzene, 83
Hexachlorethane, 83
Hugo, Victor, 41
Humbolt, Frederich Heinrich von (1769-1859), 33, 72
Huxtable, 68, 73
Hydrogen, 47

I

Inductive capacity, 78
Inductive effect, 81
Inductive electricity, 78
Innovation, 27, 65

Institute of France, 41
Intensity of light, 23
Iodine, 73
Iodine vapor, 80
Ions, 84
Iron, 83

J

Jefferson, Thomas, 23
Jenkins, M. W., 80
Jesus College, 52
Joule, James Prescott (1818-1889), 20
Juvenile lectures, 93

K

King Maximillian II, 18

L

Lamp, Miners' Safety, 58
Langmuir, Irving, 21
Laplace, Pierre Simon de (1749-1827), 57
Laughing gas, 49
Lavoisier, Antoine (1743-1794), 23, 46, 47, 55
Lavoisier, Madam, 23
Le Clanché cell, 50
Lexington, 15
Leyden Jar, 49, 67, 78

M

Macauley, 95
Magnetic field, 78
Magnetic Induction, 80
Magnetic needle, 59
Magneto Electric Generator, 82
Magrath, Mr., 68, 70
Manganese dioxide, 50
Marcet, Mrs., 67, 95

Masquerier, Mr., 68
Mayer, R. J., 20
Maximilian, Prince, 17, 18
Maxwell, James Clerk, (1831-1879), 21, 79, 100
Mechanical equivalence of heat, 20
Medical Pneumatic Institute, 48
Mercury amalgam, 54
Michelson, Albert, 21
Milan, 74
Mitchill, Dr., 48
Monosulfonic Acid, 83
Motivation, 46
Motor, 81

N

Napoleon, 23, 39, 41, 54
Napthalene, 83
Nernst, 34
Newington Surrey, 65
Nickel, 83
Nicolson, 37, 49
Nicolson's Journal, 50
Nitrous Oxide, 48, 49

O

Oersted, Hans Christian (1777-1851), 59
Oxygen, magnetic, 79

P

Paramagnetic, 81
Paris, 72, 73, 74
Park's *Chemistry*, 68
Parmentier, 19
Pasteur, Louis, 21
Peel, Sir Robert, 82
Penzance, 45
Peregini, Terresa, 39
Philosophical Magazine, 51
Phlogiston Theory, 46

Photometer, 23
Poetry, 45
Polarized light, 80
Poles, 84
Pollution, 88
Portable Gas Company, 83
Potassium, 53, 54, 69, 70
Potatoes, 19
Pupin, Michael, 79

Q

Quantitative Science, 85
Quarterly Night, The, 71
Queen Victoria, 94

R

Revolutionists, 15
Riebau, George, 66
Ritter, Johann, 37
Rolfe, Colonel, 14
Roosevelt, F. D., 16, 17
Royal Academy Woolwich, 90
Royal Dragoons, 16
Royal Institution, 3, 7, 22
Royal School Como, 39
Royal Society, 16, 20, 21, 22, 60
Royal Society Medal, 60
Rumford, Count (1753-1814), 7, 13
Rumford Historical Association, 18

S

Salem, Massachusetts, 14
Sandamanian Church, 66, 77
Scott, 95
Shakespeare, 95
Silver sulfide, 84
Smart, Mr., 76
Sodium, 53, 54, 70
Solid electrolytes, 85
Southey's Annual Anthology, 45
St. Cyr., 23

Statue of Count Rumford, 18
Statue of Sir Benjamin Thompson, 18
Steel nickel alloy, 83
Stodart, 83
Stoney, G. Johnstone, 88
Sulzer, Johann Georg, 35
Switzerland, 75

T

Tanning, 51
Tatem, Mr., 67
Taylor, 68
Terrestrial magnetism, 82
Thames River, 88
Thermo-electric, 84
Thermos bottle, 21
Thompson, Benjamin (1753-1814), 13
Thomson, Sir J. J. (1856-1940), 88
Thomson's *Chemistry*, 68
Torpedo fish, 30, 73, 92
Tories, 15
Traité Élémentaire de Chemie, 47
Transactions of the Royal Society, 83
Trinity College, Dublin, 55
Truro, 45
Tyndall, John, 21

U

University of Bologna, 30
University of Edinburgh, 46
University of Pavia, 7, 39

V

Volt, 38
Volta, Alessandro (1745-1827), 27, 72
Voltaire, 41
Voltameter, 85
Volta Pile, 34
Volta's electricity, 34
Von Helmholtz, Herman Ludwig (1821-1894), 87
Von Humbolt, Frederick Heinrich (1769-1859), 33, 72

W

Watt, Isaac, 67, 70
Watt, James, 46
Watt, Gregory, 46
Wentworth, Governor, 15
Whewell, Reverend, 85
Williamson, 88
Winthrop, Professor John, 14
Woburn, 13
Wordsworth, 61